The Politics of Aquaculture

Aquaculture is increasingly complementing global fisheries and is relevant to ocean and freshwater health, biodiversity and food security, as well as coastal management, tourism and natural heritage. This book makes the case for treating the governance of this industry as meriting attention in its own right, abandoning the polemic discussions of fish farming and opening up new ways for debating its past, present and future.

Developing and applying an original analytical framework for studying fish farming aquaculture, embedded into larger theory about the changing political system, the author generates and compares new data on the governance of aquaculture. Detailed case studies are presented of Scottish salmon, Aquitaine trout in France and seabass and seabream in Greece. The book shows how ecological issues are related to economic and social issues, as well as interdependences between territories, public and private regulation and different knowledge forms, demonstrating that these are creating alternative approaches for sustainability governance. It provides a deeper understanding of the political aspects of governing European aquaculture, including how it both is structured by and is structuring politics.

It is aimed at advanced students, researchers and professionals in aquaculture and fisheries, as well as those with a broader interest in sustainability politics and sustainability governing practices.

Caitríona Carter is a Research Professor in Political Science at ETBX, Irstea, France. She was previously Senior Lecturer in EU studies (political science) at the Europa Institute, University of Edinburgh, UK.

Routledge Studies in Environmental Policy

Public Policy and Land Exchange
Choice, Law and Praxis
Giancarlo Panagia

International Arctic Petroleum Cooperation
Barents Sea Scenarios
*Edited by Anatoli Bourmistrov, Frode Mellemvik, Alexei Bambulyak,
Ove Gudmestad, Indra Overland and Anatoly Zolotukhin*

Why REDD will Fail
Jessica L. DeShazo, Chandra Lal Pandey and Zachary A. Smith

The European Union in International Climate Change Negotiations
Stavros Afionis

The EU, US and China Tackling Climate Change
Policies and Alliances for the Anthropocene
Sophia Kalantzakos

Environmental Policy and the Pursuit of Sustainability
Chelsea Schelly and Aparajita Banerjee

Green Keynesianism and the Global Financial Crisis
Kyla Tienhaara

Governing Shale Gas
Development, Citizen Participation and Decision Making in the US, Canada,
Australia and Europe
Edited by John Whitton, Matthew Cotton, Ioan M. Charnley-Parry, Kathy Brasier

The Politics of Aquaculture
Sustainability Interdependence, Territory and Regulation in Fish Farming
Caitríona Carter

The Politics of Aquaculture

Sustainability Interdependence, Territory and Regulation in Fish Farming

Caitríona Carter

Routledge
Taylor & Francis Group
LONDON AND NEW YORK

from Routledge

First published 2018
by Routledge

2 Park Square, Milton Park, Abingdon, Oxfordshire OX14 4RN
52 Vanderbilt Avenue, New York, NY 10017

Routledge is an imprint of the Taylor & Francis Group, an informa business

First issued in paperback 2020

British Library Cataloguing-in-Publication Data
A catalogue record for this book is available from the British Library

Library of Congress Cataloging-in-Publication Data
Names: Carter, Caitríona, author.
Title: The politics of aquaculture : sustainability interdependence,
territory and regulation in fish farming / Caitríona Carter.
Description: Milton Park, Abingdon, Oxon ; New York, NY :
Routledge, 2018. | Includes bibliographical references and index.
Identifiers: LCCN 2018005518| ISBN 9781138499225 (hbk) |
ISBN 9781351014991 (ebk)
Subjects: LCSH: Aquaculture–Political aspects. | Fishery policy. |
Sustainable aquaculture.
Classification: LCC SH135 .C37 2018 | DDC 639.8–dc23
LC record available at https://lccn.loc.gov/2018005518

ISBN: 978-1-138-49922-5 (hbk)
ISBN: 978-0-367-51096-1 (pbk)

Typeset in Bembo
by Wearset Ltd, Boldon, Tyne and Wear

To Andy, Charlie, Zoé and Marnie and all our beautiful interdependencies

Contents

List of illustrations   ix
Acknowledgements   xi

1   Sustainability interdependence: aquaculture,
    sustainability, territory, regulation and knowledge   1

PART I
Theorising a politics of sustainability interdependence   29

2   Sustainability narratives of nature–society
    interdependencies and the re-organisation of state
    power   31

3   From sustainability to sustainability interdependence:
    recomposing sustainability in a new analytical
    framework   59

PART II
Institutionalising sustainability in fish farming
aquaculture   93

4   Sustainability interdependence and fish farm/
    environment interactions: governing fish farming's
    environmental impact   95

5   Sustainability interdependence and access to fish farm
    sites: environmental landscape aesthetics and coastal/
    rural development   135

6 Sustainability as a food governance problem: product
   quality and shadow ecologies   167

7 Conclusions: the 'tangled politics' of sustainability
   interdependence   207

   Annex   225
   Index   232

Illustrations

Figure

3.1 Dynamic institutionalist framework capturing political
     tensions of 'sustainability interdependence'   80

Tables

2.1 Summary of 'ideal-type' sustainability narratives   49
3.1 Research guidance and checklist   82
4.1 Summary of political work over Scottish salmon farm/
     environment interactions   107
4.2 Summary of political work over Aquitaine trout farm/
     environment interactions   116
4.3 Summary of political work over Greek seabass and
     seabream farm/environment interactions   123
5.1 Summary of political work over Scottish salmon access to
     farm sites   147
5.2 Summary of political work over Aquitaine trout access to
     farm sites   152
5.3 Summary of political work over Greek seabass and
     seabream access to farm sites   159
6.1 Summary of political work over sustainability as a food
     governance problem for Scottish salmon   183
6.2 Summary of political work over sustainability as a food
     governance problem for Aquitaine trout   191
6.3 Summary of political work over sustainability as a food
     governance problem for Greek seabass and seabream   197
7.1 Comparative contingent sustainability interdependencies of
     Scottish salmon, Aquitaine trout and Greek seabass and
     seabream   211
A1.1 Comparative sustainability interdependencies of Scottish
     salmon   226

A1.2   Comparative sustainability interdependencies of Aquitaine
        trout   227
A1.3   Comparative sustainability interdependencies of Greek
        seabass and seabream   228
A1.4   Around industry problems: farm/environment interactions   229
A1.5   Around industry problems: access to farm sites   230
A1.6   Around industry problems: farmed fish as food product   231

Acknowledgements

The research for this book was financed by the French Agence Nationale de la Recherche within the framework of the research project 'The European Government of Industry' led by Andy Smith and Bernard Jullien from 2009–2013. As I explain in the introductory chapter, field work was conducted with Clarisse Cazals and I would especially like to thank her for all our wonderful exchanges during this time. I would also particularly like to thank all those actors who agreed to be interviewed for the purposes of this research and gave up their valuable time to do so. I am extremely grateful to you all. This book has also benefitted from research carried out in the project ECOGOV, financed by the French National Research Agency (ANR) in the frame of the Investments for the Future Programme, within the Cluster of Excellence COTE (ANR-10-LABX-45). I would like to thank ECOGOV project team members for all our lively discussions on ecosystem approaches over the last few years. I would also like to thank all the members of the jury of my HDR for their constructive comments and encouragements on a previous version of this manuscript: Denis Salles, Cécile Blatrix, Anne-Cécile Douillet, Xabier Itçaina, Sylvie Ollitrault. An enormous thank you also goes to Antoine Roger, for his unstinting support throughout the process of writing this manuscript and for his valuable comments and interaction over its various chapters. I would also like to thank all my colleagues at Irstea, Bordeaux, for their thoughtful encouragement moving towards the book's completion. Finally, my warmest thanks go to my family whose support has been over and above any which I could possibly have expected, or even imagined: my thanks to you all.

1 Sustainability interdependence

Aquaculture, sustainability, territory, regulation and knowledge

Introduction

European fish farming aquaculture – the farming of fish in the sea or in freshwater – is not an industry with which most people are particularly familiar. Yet the recent growth of this industry is crystallising fundamental questions exercising society and which speak to everyday concerns. As well as ocean health, biodiversity and local food security, fish farming aquaculture touches on coastal management, tourism and natural heritage. In our changing world, this industry promises a lot. Yet this promise has been hotly contested. Indeed, although fish farming has existed for centuries in some parts of the world where its practices have been naturalised as traditional and cultural (Gibbs, 2009), its recent intensive growth following biotechnological innovation[1] has launched new debates on the purpose and outcomes of this industry. Deep tensions splice these debates: on fish farming's environmental impact, on its consequences for coastal and rural economies and natural landscapes, and on its contribution to food and fish sustainability.

In this context, a large scholarship has grown up around fish farming's general controversies and especially within the domains of ecology, fish nutrition, animal health and welfare, human health, economics and policy sciences. More recently, important contributions have also come from political economy, sociology, human geography, anthropology and law.[2] This scholarship has framed, participated in and indeed perpetuated debates on fish farming's very purpose. Strong images and ways of thinking about this industry have been built through this research, which has contributed to the analysis of numerous ecological, technological, market and policy issues linked to this industry on a global scale.

Whereas a multitude of disciplines have invested in fish farming aquaculture, this has not been the case for political science. Indeed, core political science research questions on actors and power have rarely been applied to the study of European fish farming, and the comparative politics of European fish farming has not (yet!) generated debates either in the political science community or beyond. It follows that although scholars have identified conflicts and tensions within this industry (e.g. in Norway, Tiller *et al.*, 2012; in

Canada, Noakes *et al.*, 2003); revealed the market consequences of political battles over fish farming's sustainability (e.g. in European markets, Bush and Duijf, 2011; Bush *et al.*, 2009); and pointed to the potential effectiveness of smart regulatory policy tools (e.g. in Canada, Howlett and Rayner, 2004); there has been little comparative analysis either of actual governing practices or of policy directions chosen in different European territories. As a result, we lack insight on the deeply political aspects of governing European fish farming aquaculture, including on how European fish farming is both structured by and structuring of politics.

An underlying aim of this book is to respond to this challenge. Rather than just assuming that the politics of fish farming aquaculture is 'well known',[3] the book makes the case for treating the governance of this industry as meriting attention in its own right. Developing and applying an original analytical framework of 'sustainability interdependence', the book generates new data on the governing of Scottish salmon, Aquitaine trout and Greek seabass and seabream in relation to three sustainability challenges: (i) interactions between fish farming and the environment; (ii) inter-relations between the continued growth of European fish farming and rural and coastal territorial development; and (iii) farmed fish as sustainable food.

Throughout, a particular aim of this book is to reappraise sustainability governance through emphasising what I have called 'sustainability interdependence'. Sustainability is of course a widely analysed concept which at its base recognises important interdependencies between society and nature (Hopwood *et al.*, 2005: 40).[4] Yet for some scholars, tensions between meeting societal goals and preventing environmental degradation have become so great that they regard sustainability to be obsolete as a concept and a policy programme (Benson and Craig, 2014). Although this is not the starting point of the book, which recognises that sustainability continues to shape actors and their action, nevertheless I agree that our analysis of sustainability requires reappraisal in light of the complexity of current environmental challenges. More specifically, drawing upon long-standing premises of public policy analysis highlighting the importance of policy–polity tensions in public action transformation, the book argues that sustainability requires both updating as a concept and recomposing as part of a broader theory of political change.

To meet these objectives, the book builds an original analytical model of 'sustainability interdependence'. The book argues that in political struggles over sustainability, ecological issues are not only brought into relation with economic and social ones (classic sustainable development), but also into relation with other interdependencies on-going in the political system. These are (i) interdependencies between territories (brought about through processes of globalisation and regionalisation); (ii) interdependencies between public and private regulation (following from changing relations between states and markets); and (iii) interdependencies between different knowledge forms (resulting from the democratisation of knowledge use in decision-making). Moreover, it is through seizing these broader interdependencies that actors have managed to open up new spaces for putting in place alternative sustainability governing approaches – and in so doing alter fish farming's relationship to the environment. Examining the governing of fish farming through the lens of sustainability interdependence therefore allows for the development of a nuanced critique of its governing practices, offering new understandings of both successes and failures in governance. In so doing, the book abandons polemical discussions on fish farming aquaculture – either 'for' or 'against' it – and opens up new ways for debating this industry's past, present and future.

In this introductory chapter, I first set out the origins of this argument and then go on to discuss my selection of case studies, methods and data. I conclude with some general remarks on the overall positioning of the book.

Conceptualising 'sustainability' in fish farming aquaculture

The approach adopted in this book has its origins in the consideration of two relationships pertinent to a political analysis of European fish farming aquaculture. These are first, the relationship between aquaculture and sustainability, and second, the relationship between choices made over sustainability in local governing practices and broader processes of transformation of the political system. In the next two sections I wish to outline the origins of this approach and sketch out the main elements of conceptual development which it entails – and which are developed further in Chapters 2 and 3.

A central contribution of this book is to propose a renewal of the concept of sustainability through integrating it into a wider inquiry on the politics of interdependence. Before setting out how this will be carried out, I begin with a discussion on the more general relationship between aquaculture and sustainability. This is because the empirical starting point of this book is that a primary struggle on sustainability lies at the heart of the European fish farming industry (Carter, 2012).[5] In European fish farming, the debate on its sustainability and how its various economic, ecological and social challenges are defined and brought together is a primordial one, turning on the industry's very identity and the justification (or not) of its existence. This contrasts fish farming aquaculture with other industries, for example sea fisheries, where a relative consensus exists both on fisheries' contribution to society's industrial heritage and on what constitutes unsustainable versus sustainable fisheries practice.[6] In fish farming aquaculture, extreme positions are taken up: fish farming is either sustainable *by its very nature*, or *its very existence* as an industry *is called into question*. Mixed positionings are additionally taken up by numerous groups in between these two extremes.

Some examples will illustrate this important characteristic of this industry. For example, early claims legitimising the intensive development of European fish farming in the 1980s and 1990s argued that the development of this industry was necessary to secure the sustainability of fish resources. It was argued that the very existence of fish farming would reduce pressures on collapsing wild capture fisheries by supplementing and substituting caught fish with farmed fish. This was a powerful argument justifying support for this industry (Costa-Pierce, 2010). Public actors have described this as follows:

> [T]he conventional school of thought is that we are overfishing our capture fisheries and aquaculture offers a way forward for supplying alternative protein and we ought to be growing it.
>
> (Interview: Scottish regulatory agency official, 2010)

Of course, this is by no means to imply that fish farming was considered by these groups to be a conservation industry.[7] Rather, the business opportunities which it offered were presented also in relation to the sustainability of wild fish populations:

> We are not doing this as a conservation project. There is a declining supply of tuna. With declining supply comes increasing prices. So there is an opportunity to support that supply and take the pressure off the wild fishery by producing it in a farmed manner. Hopefully, if we can succeed, then we will take pressure of fishing stocks and the world will go on as we would have found it years ago.
>
> (CEO of Clean Seas Tuna Research Ltd., speaking with Simon Reeve in the BBC documentary on Australia, 2014)[8]

These arguments favouring aquaculture in the name of the sustainability of wild resources (and seafood markets) have also been linked to others concerning food security in response to world population growth and ever-increasing protein needs. Accordingly, farm-fed fish can be viewed as an 'efficient mass producer of animal protein' (Costa-Pierce, 2010: 96), superior in terms of energy and production efficiencies compared with both capture fisheries and terrestrial agriculture systems (Costa-Pierce, 2008: 317). Growing fish fed on other fish which are not fit for human consumption is presented as the most effective way to convert otherwise inedible marine proteins and oils into food – as distinct from using these marine resources to feed pigs or cattle where they are simply wasted (Bureau, 2006). To quote Simon Reeve's comments following his visit to Clean Seas Tuna Research Ltd.:

> Of course, it is sad to see such magnificent creatures being held captive like this and farmed, but we have been doing the same to cattle for thousands of years and at the moment the human population of the planet is increasing by tens of millions every year and we are emptying our oceans of fish … maybe fish farming – aquaculture – can play a role in finding a solution which feeds humans but protects life in our seas.
>
> (Simon Reeve, BBC documentary on Australia, 2014)[9]

This vision of aquaculture as promoting sustainability is not, however, shared by all. This is because it is argued that intensive industrial fish farming has

brought with it its own local negative environmental impact, including negative interactions with wild fish populations following from fish escapees, the
spread of disease and fish health problems, and water pollution from organic
matter and chemical treatments (Porter, 2005; Russell *et al.*, 2011). Farming
fish has also been argued to have a direct negative impact on in-shore fish
species, where controversies have been documented between fishers and fish
farmers over the negative impacts of salmon farming (Wiber *et al.*, 2012).
Aquaculture's role as an efficient converter of marine proteins has also been
the topic of fierce international debates where leading scientists have argued
that the farming of fish is putting strain on feed fisheries and thereby contributing to the *collapse* of wild fish stocks – as distinct from the opposite (Naylor
et al., 2000). In this situation:

> [A]quaculture's blue revolution suddenly seems less promising. Aquacul-
> ture's 'innovation' may be associated with the introduction of new,
> undesirable risks.
>
> (Castle and Culver, 2008: 2)

Consequently, whereas for some, fish farming aquaculture can be a sustainable solution to collapsing wild fish stocks and demands for food security, for
others, it can be an important cause of aquatic ecosystem degradation. Arguments along these lines have been mobilised in particular by those NGOs
which are outright hostile to intensive industrial fish farming. These NGOs
have argued that technological innovation has merely replaced one set of ecological problems linked to fisheries with another set of ecological problems
linked to fish farming – and has not necessarily 'solved' the problem of fisheries along the way (Huntington, 2004).

Another example of fish farming aquaculture's contested relationship with
sustainability is over its role in rural territorial development. This turns on
debates around potentially competing forms of rural economy – the blue rural
economy versus the 'retirement/recreation/rural home business economy'
(Howlett and Rayner, 2004: 182). Aquaculture has been argued to be a
pivotal industry for local territorial development, especially in rural and
coastal areas where its presence can grow local economies, thus contributing
to the protection of the social fabric of (coastal and inland) rural life and social
territorial equality. New employment opportunities can provide stability
when other industries are in decline (Rana, 2007). For example, 'in Scotland,
the salmon industry is regarded as the most important economic development
in the Highlands and Islands' (Rana, 2007: 44). Yet, whereas in some places
fish farming's contribution to local employment is valorised, in others it has
been presented as working against local development of a different kind,
namely coastal tourism. In the face of demographic change and the consolida-
tion of the coastal tourist industry, fish farming has new opponents in the
form of second homers, retirees and hotel owners who claim that aquaculture
is ruining the natural coastline:

[Y]ou cannot grow any fish in the southern part of France now, in Corsica, no way. It is not because it is full, but because nobody wants to see it ... people go to the seaside: they don't want to see farming activity in the sea.

(Interview: European Commission DG Research public official, 2010)

Polarising relationships over fish farming aquaculture and sustainability can also be found in discussions on consumer health. In 2004, a scientific article published in *Science* sparked a fierce debate in North America and the European Union (EU) over human health impacts of eating farmed Scottish salmon (Hites *et al.*, 2004). By contrast, more recently the debate has shifted towards the positive contribution which finfish aquaculture can make to mitigating mental health and brain disorders, e.g. Alzheimer's. Human nutrition experts have argued that, in a world where human mental health disorders are increasing, causing personal stress and imposing unsustainable costs on health systems, fish farming aquaculture can contribute to the sustainability of human mental health. This is because certain species of farmed fish (e.g. salmon or trout) can be efficient converters of the fatty acids Omega 3 and 6 contained in the fish oils in their diets and which are essential for protecting human mental health (Crawford, 2010).

It thus becomes apparent that any analysis of fish farming aquaculture must address its relationship to sustainability. Indeed, scientific research associated with this industry engages directly with this relationship. For example, scholarship emerging around this industry has discussed fish farming aquaculture's negative environmental, territorial or human health impact (Hites *et al.*, 2004; Naylor *et al.*, 2000;); presented results demonstrating positive and improving farm-environment interactions (Costa-Pierce, 2010; Soto *et al.*, 2008; Tacon *et al.*, 2010); and proposed solutions to reduce or mitigate negative impacts (Agúndez *et al.*, 2013; Grant and Treasurer, 1993).

In this light, when so much has already been said and done on fish farming's relationship to sustainability, what new line of inquiry can this book open up? In response, and through drawing on political science's core research questions around actors and power in policy-making, the book proposes to take as its focus *the governing of* aquaculture's relationship to sustainability. How do those governing this industry understand its relationship to sustainability? Which governing choices are made to restructure this relationship? Which conflicts have taken place over this very question and how have these been resolved? Built into this focus is an assumption that aquaculture's relationship to sustainability is neither inevitable nor spontaneous, but can be changed through the act of governing (Carter *et al.*, 2014a; Jullien and Smith, 2014). The book explores how different sustainability problems facing this industry have been rendered 'governable' over time, comparing sectors and territories. Indeed, as the empirical chapters of this book reveal, through governing, fish farming's sustainability can take on a variety of forms and meanings. Furthermore, these governing choices are more or less stabilised in

regard to Scottish salmon versus Aquitaine trout versus Greek seabass and seabream compared.

Of course, the importance of an analytical focus on governing in studies of
sustainability has been promoted for decades within an extensive scholarship
in the social sciences (e.g. Jordan and Adelle, 2013; Lafferty and Hovden,
2003).[10] In this vein (and as I argue in more detail in Chapters 2 and 3), a
strong tendency has been to study sustainability prescriptively, define it as a
specific policy goal and assess its government in terms of its ability to meet
that goal:

> [O]ne must believe that the task of achieving sustainable development is
> a rational one: a process that can, to a reasonable degree, be 'steered' by
> governing procedures and institutions.
>
> (Lafferty, 2004: 3–4)

Whilst I share with this work the conviction that understanding governing
practices and processes of policy decision are critical for understanding relationships of industrial production and sustainability, nevertheless in this book
I approach this question from a different ontological angle. More specifically,
drawing on 'inclusive' (Kauppi, 2010) and inter-subjective conceptions of
institutions defined as norms and rules, the first proposal in this book is to
treat sustainability heuristically as an institution – by which I mean as a public
action principle – governing this industry. To be clear, this is neither to con-
ceive of sustainability in rationalist institutional terms as a goal (Lafferty, 2004:
3–4), nor to consider that full consensus necessarily exists on the contents of
this public action principle (Mouffe, 2000). On the contrary, fish farming's
sustainability is the discussion's main cleavage, polarising debates between
groups of proponents and opponents (Castle and Culver, 2008: 2). What is
meant, rather, is that sustainability is seen by the actors themselves as *unavoid-
able institutional practice* in today's governing of European fish farming
aquaculture:

> When I talk about aquaculture I always talk about sustainable aquaculture
> because this is our starting point.
>
> (Interview: European Commission DG MARE public official, 2013)

> It is living and breathing it – it is not just bolt on – 'oh we need to think
> about that'. Everybody knows that, so it is living and breathing it. And it
> is not just me, you know going out to buyers or whatever. They all
> know. It is embedded. It is a fundamental within the business.
>
> (Interview: UK supermarket, 2012)

Paraphrasing Pierre François, this is to understand that the power of the institution of sustainability does not stem from any intrinsic characteristics of this
public action principle, but from the fact that actors across this industry give

it this role and allow it to intervene in their practices as an a priori and perennial frame (François, 2011: 43).

This way of thinking about sustainability thus differs from it being considered either as a linear process to reach some clearly defined goal (Lange *et al.*, 2013) or as obsolete (Benson and Craig, 2014). Rather, the building of sustainability as an institution is understood here to be a 'contested', 'constantly emergent' and 'relational' practice (paraphrasing from Healey's analysis of territorial development; 2004: 48–50). This is because institutions are not 'things [but are] socially embedded processes … and what people do' (Cleaver and Franks, 2005: 3). As argued further in Chapter 3, institutional politics is viewed as seeking to achieve 'unity in a context of conflict and diversity' (Mouffe, 2000: 15; Lawrence and Suddaby, 2006: 248), but where there will always be exclusion because the achievement of a 'fully inclusive rational consensus' is neither possible nor at issue (Mouffe, 2000: 6).

Sustainability's respective contents thus represent 'settlements of conflicts' (Bartley, 2007), whereby dominant actors govern in a situation of tension and alternatives (Schneiberg, 2007). This all comes about through 'institutional work' (Lawrence and Suddaby, 2006) and 'political work' (Jullien and Smith, 2014). In this manner, sustainability's various meanings and instruments are understood to be created, maintained and/or disrupted by groups of actors over time (Lawrence and Suddaby, 2006); they are institutionalised, de-institutionalised and re-institutionalised (Jullien and Smith, 2014). In this way, sustainability's uptake in governing practices can be grasped as a temporal process studied over long periods of time (i.e. over decades). Additionally, in these institutional processes characterised by diverse trajectories in different locations, sustainability as a public action principle not only offers actors resources in the form of narratives and theories about economical, ecological and social interdependencies and values (see Chapter 2), but simultaneously serves as a 'vector of power', around which identities and authorities are built (Merand, 2008).

In summary, first I propose resolving a concern over how to study the relationship between aquaculture and sustainability through centring the book's analytical focus on the governing of aquaculture's sustainability. Second, through drawing on the well-honed public policy concepts of institutions and institutional and political work, I propose defining sustainability as a public action principle whose creation, stabilisation, contestation and re-creation can be compared in different sectors and territories within European fish farming.

Integrating 'sustainability' into a broader framework of 'sustainability interdependence'

This brings us to the second relationship pertinent to a political analysis of fish farming aquaculture. As I state above, a central objective of this book is to recompose our very way of thinking about sustainability through integrating it into a broader politics of interdependence. To build our approach towards this objective, we must consider the relationship between choices made over sustainability in local governing practices and broader processes of transformation of the political system. Indeed, as we know from long-standing research within public policy analysis, the resolution of an industry's sectoral problems is not a process closed off from global transformation (Jobert and Muller, 1987; Muller, 2015). Industries are brought into relation with society through the 'social, political, contextual, historical construction of economic activity' (Jullien and Smith, 2014: 4), whereby sectoral choices are shaped by larger concerns. Through governing sectors, society both acts on itself and enacts its response to global processes (Muller, 2010; 2015).

In particular, the institutionalisation of sustainability in the governance of fish farming aquaculture takes place in a wider political and economic environment of global and climate change. As a broad international scholarship can attest, this is a dynamic environment in which the political organisation of the liberal capitalist state is a constant source for debate and analysis. In this regard, we can identify three sets of changing regularities in political power which have animated international scholarly discussions over the past decades: (i) the territorialisation of politics and the changing role of the Westphalian state; (ii) changing relations between states and markets and transfers of political regulatory authority from the public to the private realm; and (iii) the democratisation of knowledge use in policy-making. As becomes apparent from a rich body of work describing these changes and stretching over many years (which is discussed in detail in Chapter 3), changing regularities within the political system have given rise to three types of interdependence in addition to those traditionally associated with sustainability. These are (i) interdependence and territory; (ii) interdependence and regulation; and (iii) interdependence and knowledge.

The emergence of these interdependencies can represent 'a major new – or renewed – challenge to the theory and practice of politics in Europe' (Jeffery and Wincott, 2010: 168). First, the re-organisation of Westphalian state power in Europe, both through processes of European integration as well as processes of decentralisation and devolution of state sovereignty and politics, has re-dispersed political power both above and across the nation state. This has recreated new political communities and identities, new public actors and new public policy resources at different scales and hence a range of political territories with emerging (and potentially conflictual) territorial interests.

Second, state projects of neoliberalisation along with public sector and regulatory reform have altered the ways in which states interact with markets. Within these changing relationships, a significant development has been the redistribution of power from public to private regulators and the shift in authority over regulation from the state to civil society. This has resulted in the development of new forms of co-regulation and private self-regulation (and new economic 'governors', at times regulating with NGOs), which exist alongside public instruments of governance and public actor governors.

Third, the combined diminished authority of linear science–politics relations and the rise in participatory governance and the democratisation of knowledge have created a new era of multiple knowledge use in policy-making. Myriad knowledge forms can and have been mobilised in policy-making, including, *inter alia*, regulatory science, usable science, expertise, technical expertise, user knowledge, everyday knowledge, traditional knowledge, citizen science and local knowledge. This too has given rise to a range of new actors in policy-making processes. On-going change has thus generated new sets of interdependence which have produced their own conflicts, contradictions and tensions – both in scientific analysis and in governing practice – and especially over critical societal issues of efficiency, legitimacy and democracy in governance.

Rather than treating these interdependencies as the background political context in which this study takes place, the second proposal in this book is to treat them instead as a key object of research. This has certain implications for research because the very idea of 'interdependence' points to a specific understanding of relationships.[11] According to Hay, interdependence exists when change in one component or variable will result in change in another (2010: 6). Importantly, there must be no consistent causal direction between the two components, which is rather a relationship of dependence, not interdependence: 'A' can equally affect change in 'B' and 'C', as 'C' can affect change in 'A' and 'B' (Hay, 2010: 7). Hay argues further that whereas 'interdependence is a neutral and descriptive concept – neither innately good nor innately bad', it is not 'politically neutral' and is 'difficult to govern' (2010: 7). These underlying conceptions of interdependence lie at the heart of my approach. Consequently, studying interdependencies necessarily involves studying dynamic tensions between the different components which make up the interdependence and, more specifically, grasping how actors mediate these tensions when governing socio-ecological problems – what I term the 'frontier politics' of interdependence.

Thinking in terms of broader interdependencies also holds implications for our understanding of the very purpose of 'frontier politics'. Taking inspiration from political sociology, which defines politics as 'action which seeks to change the distribution of power' (Frickel and Moore 2006: 10), institutional and political work undertaken by actors to govern fish farming aquaculture's problems is not only conceptualised as a political struggle over the substance of sustainability (and resultant norms and rules), but also over the very authority to set these norms and rules and govern relations in the first place, i.e. the authority to influence (Genieys and Smyrl, 2008). In this vein, territorial, regulatory and knowledge interdependencies provide additional potential political resources and potential 'collective forms of identification' (Mouffe 2000) for actors beyond social, economic and ecological theories, arguments and values. Conceived in this way, we can ask to what extent the various groups of actors governing fish farming aquaculture are active users of these material and imagined resources in their institutional and political work over sustainability problems. For as argued by Lawrence and Suddaby, institutional work require[s] resources, which are available to some actors and not others. A critical view of institutional work could begin to examine how those resources are distributed and controlled, and by whom.

(2006: 247)

This gives rise to two sets of research questions addressed in this book:

i   How does the politics of interdependencies of territory, regulation and knowledge ultimately come to shape the way in which sustainability problems are rendered governable? For example, do actors mobilise around the frontier politics of these interdependencies to create contingency and open space for alternative policy directions in the governing of sustainability? Or do they rather experience any contradictions contained therein as constraints?
ii How does the institutionalisation of sustainability as a governing problem for European fish farming aquaculture in turn affect broader processes of transformation of the political system? How does the institutional and political work of actors in European fish farming aquaculture contribute to the perpetuation of these interdependencies?

In summary, the solution proposed in the book for addressing the second relationship between choices made over sustainability in local governing practices and broader processes of transformation of the political system is to embed an institutionalist analysis of sustainability into larger theories of the changing political system. The approach to European fish farming aquaculture is not therefore limited to an analysis of how local industry conflicts are shaped and governed in different sectors and territories – although this remains a central question – but rather expands to consideration of the broader political consequences of the local policy directions chosen.

This creates a challenge for research. As I explain in Chapters 2 and 3, changes to societal democracy and changes in the political system as defined above have also been heavily influenced by 'sustainability narratives' (ecological modernisation, sustainable development, eco-democracy, ecosystem approaches and sustainability as de-growth: see Chapter 2). These ecologically induced counter developments have given rise to political projects which can also be viewed as important drivers of processes of change (Eckersley 2004), contributing both to the politics of interdependence and territory, regulation and knowledge, and to how we study this politics (Chapter 3). Yet, as Chapter 3 also argues, these are not the only (or necessarily key) drivers of change and this thus gives rise to what I ultimately refer to as 'a tangled politics of sustainability interdependence'.

Case studies

Whilst the first general objective of the book is to develop an analytical approach embedded in a larger theory about actor strategies within a changing political system, the second general objective is to apply this approach to empirical case studies relevant to European fish farming in different sectors and territories, namely: Scottish salmon, Aquitaine trout and Greek seabass and seabream. In general terms, the case studies (material for which was initially gathered working with Clarisse Cazals from Irstea, Bordeaux) make it possible both to localise the inquiry around the governance of fish farming aquaculture and to produce new knowledge on actual governing practices, addressing political controversies over sustainability in different settings. Hence, the method in the case-study design has been problem-orientated rather than country-orientated.[12] Organising each case around a sustainability problem facing the industry, I analyse how this problem is governed in respect of Scottish salmon, Aquitaine trout and Greek seabass and seabream compared. This includes comparing actors' institutional and political work and answering the research questions as set out above on the role of the politics of interdependencies in this field.

Before setting out the problems around which our case studies have been organised, it is important to provide some background information concerning the three sectors and territories chosen, namely: Scottish salmon, Aquitaine trout and Greek seabass and seabream.

Salmon, trout, seabass and seabream are considered important fish farming aquaculture sectors in Europe, representing the top four EU farmed fish products.[13] Defined by the European Commission as a critical industry for the EU bio-economy, intensive European fish farming has developed following biotechnological innovation in the 1970s and 1980s. All four species of fish have been grown in this way. For example, salmon farming consists of freshwater hatcheries and the raising of salmon in net cages in freshwater and seawater lochs on the West Coast, Highlands and Islands, Orkney and Shetland (Carter and Cazals, 2014). In Scotland it began initially as a small-scale activity in the 1970s (the first salmon produced in 1971 by Unilever – now Marine Harvest) to diversify crofting, which is a traditional Scottish form of community-based small-scale food production. Individual land farmers experimented with fish farming applying new biophysical technologies and soon realised the potential for expansion, leading to the separating out of a distinct industry in the 1980s.

In Aquitaine, trout farming also began in the 1950s as small family-owned businesses – some of which still exist today (e.g. La Truite des Pyrenées) – although not in the sea but along rivers originating in the Pyrenean mountains. Trout farming differs from salmon farming as it takes place next to a river, whereby a portion of the water flow from the river is diverted through long tanks of fish which make up the farm site (Carter and Cazals, 2014). In the 1980s, the Landes region of Aquitaine was experiencing serious economic decline with the collapse of its resin industry. Understanding that there was an untapped potential of river water resources in the region, a small group of newly established fish farmers came together in 1981 to create a fish farm cooperative (today Aqualande) and experimented with innovative biotechnological techniques to grow larger trout.

In Greece, the farming of seabass and seabream began in 1981. Like salmon, seabass and seabream are raised in net cages in in-shore marine waters, although unlike salmon they are initially grown in seawater hatcheries. As is the case for the other two industries, early beginnings saw the emergence of many small farms, pioneering biotechnological innovation often adapted from salmon farming and through the employment of scientists from Northern Europe to transfer know-how (in particular pioneering hatcheries).

Since these early beginnings, all three sectors initially experienced similar high rates of growth, but this soon changed as the growth of the trout industry began to stabilise and even decline, with production levels reducing overall since the 1990s. According to the European Commission, it is salmon, seabass and seabream which are currently 'the only species with a consistent growth in value and volume' (European Commission 2009a: 66). In Scotland, this growth history has been marked by three different periods of expansion (c.1985–1995; c.1995–2002–2003; 2003 to present), and two crashes (1993–1995; 2002–2003). In Greece, early growth was rapid. For example, between 1990 and 1999, the average annual increase in production was measured at 70 per cent (European Commission, 2009b: 64). The industry has continued to expand since, with two general market crises in 2002–2004 and 2007–2010.

Despite differences in growth rates, in all three cases growth rates do not match EU consumer demand for fish and seafood products. Actually this demand is increasingly being met by seafood imports (65 per cent of EU consumers' seafood needs are met by imports), much of which stem from aquaculture production elsewhere in the world (worldwide output of aquaculture production has doubled in the past ten years: Castle and Culver, 2008: 93). Against this background, an important feature of these sectors is the conviction which many industry and public actors hold over their *potential* for growth in Europe. This is to argue that fish farming could be a key industry for the future, critical for European sustainable food security and employment in coastal and rural areas (although, as we shall see, this growth project is highly contested).

The company profile of these sectors is also quite specific to fish farming aquaculture, as distinct from other forms of aquaculture (e.g. shellfisheries, such as mussels and oysters). After having experienced early boom and bust growth patterns, contraction coupled with company buy-outs and mergers has resulted in the establishment of large fish farming companies, alongside some remaining family-run businesses. Indeed, although the EU aquaculture industry in general (including shellfish farming) only has a small number of large and medium-sized enterprises (2 per cent have a turnover exceeding

20m€), thirteen out of fifteen of these firms are producers of marine fish farming aquaculture (Ernst & Young Report 2008). For example, in Scotland between 2002 and 2010, the number of companies producing salmon was reduced from 55 to 31 (Scottish Government, 2010: 14).

During our research period (2009–2013), four large listed companies (Marine Harvest, Scottish Sea Farms, Grieg Seafood Hjaltland and Meridian) represented almost 90 per cent of Scottish production. In Greece, whereas there were around 150 firms operating in the market in the 1990s, production is today dominated by a handful of large multinationals (e.g. Selonda, Nireus, Dias), eight of which are publicly listed companies. Whilst in Scotland, it is a case of Norwegian companies entering the market, in Greece it is one of Greek companies working on the stock market to stabilise their financial position. In Aquitaine, the cooperative Aqualande has grown over the years from 5 to 29 members but has now reached its point of stability. During the research period, Aquitaine production was dominated by two large firms, Aqualande (a cooperative) and Viviers de France (an international group).

These three sectors also have in common challenges concerning their territorial integration. The growth of intensive fish farming has incurred political opposition in the form of both NGO social movement activism against fish farming and community resistance to its development. These controversies are a constant feature of this industry and turn on a range of complicated factors which require unpacking. For example, accusations have been made against fish farming in all three sectors on the grounds of its environmental impact. However, this industry has not developed in an unregulated fashion. On the contrary, all three sectors are heavily governed by extensive EU regulation. Although no common EU aquaculture policy exists, for example along the lines of the EU Common Agriculture Policy,[14] nevertheless general EU regulations on environmental protection and biodiversity, food safety, and fish and animal health govern the majority of production practices in all three sectors and territories (Carter and Cazals, 2014).

Another ambiguity exists over fish farming's relationship to the local communities in which it has become established. In order to compete and grow, technological innovation has become a feature of this industry; but technological innovation can create its own set of problems. It can contribute to critiques against the 'naturalness' of fish farming, both biophysically and also in relation to local ecosystems and landscapes. It can also fuel an image of this industry as one which has 'come in' from the outside (compared to traditional fisheries, for example) and therefore is not part of industrial heritage. Both of these issues play out in local community politics. Finally, all of these relationships have to be squared with high consumer demand for seafood products. Indeed, there often appears to be a mismatch between consumer demand for farmed fish products versus residents' acceptance of fish farming production.

Against this general background, towards the middle of the 2000s, these different sectors and territories were facing three sustainability governing challenges which they held in common: first, sustainability and the challenge

---

Box 1.1 Key EU regulation potentially governing aquaculture

EC Species and Habitats Directive 92/43/EEC

Environmental Impact Assessment 85/11/EEC as amended by 97/11/EC (Aquaculture mentioned in Annex II)

Water Framework Directive 2000/60/EC

EC Strategic Environmental Assessment Directive 2001/42/EC

EU Fish Health Directive 2006/88/EC

EU Marine Strategy Framework Directive 2008/56/EC

EU Birds Directive 2009/147/EC

Rules on Common Market Organisation within the reformed Common Fisheries Policy 2013

Maritime Spatial Planning Directive 2014/89/EU

European Maritime and Fisheries Fund Regulation 2014/508/EC

EC directives and regulations on Animal Transportation

EC directives and regulations on the marketing of Veterinary Medicines

EC regulations and decisions on Food Safety

EC rules on Integrated Coastal Zone Management and Integrated Maritime Policy

EU Directive on Discharges 2000/532/EC

EU Directive on Environmental Responsibility 2004/35/EC

---

of fish farm/environment interactions (Chapter 4); second, sustainability and the challenge of access to farm sites (Chapter 5); third, sustainability and the challenge of farmed fish as food (Chapter 6). These problems were identified by EU-wide stakeholders as fundamental ones during political engagement in the development of the European strategy on sustainable development for aquaculture 2009 (Carter and Cazals, 2014: Carter *et al.*, 2014a) and consistently emerged as such in interviews with actors in Scotland, Aquitaine, Greece and in Brussels (see below on methods and data).

The first problem – *sustainability as a 'fish farm/environment interaction' governing problem* – concerns the governing of the environmental impact of fish farms, for example of farm sites along river banks (in the case of trout farming) or of floating net cages in freshwater and seawater lochs (in the case of salmon, seabass and seabream farming). As discussed above, fish farm aquaculture/environment interactions can pertain *inter alia* to environmental issues of water quality, environmental impacts on the seabed and interactions with wild species including the spread of diseases. The issue of aquaculture environment interactions has been a major one facing the industry and a source of controversy and conflict, whereby environmental NGOs with fish farming-specific remits have been created with the express intention of attacking fish farming on this issue (through mediatising environmental impacts, taking court action etc.).[15] Its importance for science is such that a new scientific journal has been created on this very theme (*Aquaculture Environment Interactions*). In keeping with my line of inquiry in the book, rather than viewing these interactions as spontaneous (Jullien and Smith, 2014), my aim is to demonstrate how these have been governed over time (see Chapter 4). This reveals a tangled politics of sustainability interdependence over both local and wider ecosystem impacts.

The second problem – *sustainability as an 'access to fish farm sites' governing problem* – concerns the growth of the industry in particular locations, for example along rivers in rural and mountain settings (trout farming) or in near shore or exposed marine locations (salmon, seabass and seabream farming). The sectors I explore were newly created in those territories where, since their early establishment, they have faced important challenges. These include becoming territorially embedded in the regions where they have developed, where it is not certain that they have succeeded (see Chapter 5). Having access to farm sites is linked to broader local territorial development strategies and how societies imagine river and coastal futures. This issue thus raises questions over the equality of nature's benefits to society (questions around 'who' should benefit from nature's resources) and, as I will show in Chapter 5, is highly controversial. In these debates, fish farming is often pitched against tourism (which carries its own sustainability challenges).

The third problem – *sustainability as a 'food' governance problem for farmed fish products* – concerns issues of food safety and food quality and the all-important question of fish feeds. Farmed fish must be fed and feeds are the most important economic cost of a fish farm's business. For example, feed costs represent 46 per cent of total operating costs for salmon; and 54 per cent of the costs of sea bream (European Commission 2009a: 135). Initially, farmed fish food products were deemed sustainable by their very nature: biotechnological innovation could reduce pressures on collapsing wild capture fisheries by replacing wild fish with farmed fish in the market place. This narrative of fish farming as 'naturally sustainable' has since been contested through political action against this industry. In Scotland and in Aquitaine, producers have responded by building alliances with NGOs and supermarkets to self-regulate and set private sustainability standards for farmed fish products. By contrast, in Greece, there has been no governing response: whereas smaller firms sell organic products into niche markets, dominant actors continue to sell onto a generic market. As I will show in Chapter 6, political controversies over farmed fish as food extend over wide spatial scales, from the local up to the global.

In summary, these empirical chapters aim to present new, relevant and otherwise difficult to find knowledge about fish farming governance choices and policies. In so doing, they can also provide insights into societal responses to major challenges such as ocean health, biodiversity and local food security.

Methods and data

In my analysis of the way in which the different problems outlined above have been worked on by actors in the different sectors and territories concerned, I mobilise empirical material which was initially gathered working with Clarisse Cazals (Irstea, Bordeaux) within the framework of a larger research project on 'The European Government of Industry' funded by the French national research agency (ANR) under the co-leadership of Bernard Jullien (GREThA, Bordeaux) and Andy Smith (Sciences Po, Bordeaux). In this ANR project, different kinds of qualitative data were gathered on the European government of fish farming aquaculture over a four-year period from 2009 to 2013.

Research activities to gather these data included document and scientific article collection, situated observation of practitioner and stakeholder meetings, and over 60 semi-structured elite interviews (lasting one-and-a-half to two hours) conducted in Scotland, France, Greece and Brussels with a full range of actors across this industry. This included interviews with actors from the European Commission, the European Parliament, the Committee of the Regions, the Economic and Social Committee, the UK representation in Brussels, Scottish government departments and its environmental and heritage agencies, French government ministries, the Aquitaine Regional Council, the Greek government, fish farming companies, feed manufacturers, inter-professional organisations of producers, inter-professional organisations of feed manufacturers, inter-professional organisations of fish processors, retailers, scientists, collective research organisations, and consumer and environmental NGOs (e.g. WWF, MCS, BEUC).

The ANR project was informed by an interdisciplinary research framework bridging industrial economics with a political sociology of institutions and applying a common line of questioning across different industries concerning their European government (wine, automobiles, pharmaceuticals and aquaculture). Some of the data gathered on aquaculture were subsequently analysed in order to produce results responding to the project's research questions and more precisely to substantiate project arguments on the nature of the European government of industry as being 'omnipresent, yet incomplete and largely depoliticised' (Carter, 2012, 2015; Carter and Cazals, 2014; Carter *et al.*, 2014a, 2014b; Jullien and Smith, 2014: 21).

The book builds on this work, going further to produce a comprehensive analysis of all the data gathered applying the analytical framework of sustainability interdependence. To this end, I have analysed different categories of documents including: position papers, commissioned reports and policy documents of public bodies, collective private organisations and NGOs; stakeholder responses to European Commission consultations; speeches at conferences and to the press; stakeholder practitioner meetings observed. The interview transcripts were all uploaded on NViVo 8 software which was used to manage the data for coding and query purposes. Transcripts were coded 'line by line and paragraph by paragraph' (Huising and Silbey, 2011: 23) and their contents categorised both with respect to the case studies chosen and concerning the different categories of the analytical grid (see Chapter 3). As explained by Young *et al.*:

> [C]oding means carefully reading and demarcating sections of the data
> according to what they represent: each code represents one concept, and
> multiple codes can be applied to one piece of data … complex queries
> [can] be performed to explore relationships between concepts, thus aiding
> the researcher to comprehensively explore and interrogate patterns within
> the data.
>
> (2014: 391)

Through coding thematically and inductively on the politics of the different
interdependencies, I was able to code different pieces of text multiple ways. I
then ran coding queries to produce different combinations of findings and I
worked back and forth in an iterative way (Young *et al.*, 2014). I was also
able to generate visual tables and charts of my data, which helped me in my
analysis, including confirming the omnipresence of the sustainability politics
of territorial, regulatory and knowledge interdependencies across case studies
and territories.

In keeping with the book's ontological approach to the study of institutional and political work (see Chapter 3), I first analysed the qualitative data
to identify changing trajectories of governing the sustainability of fish farming
over time (i.e. over decades). Operating at a meso level of analysis, I have
thereby been able to provide synthetic temporal accounts of the political and
institutional work of actors on farm/environment interactions, access to farm
sites and farmed fish as sustainable food, showing changes in meanings of
sustainability governance over time. I have not limited my analysis of the data
for purposes of historicity, however, but have also sought comprehension of
actors' representations of events (Pinson and Sala Pala, 2007).

Throughout, my aim has been to go 'inside' governing practices and the
institutional and political work of actors in the three case studies explored.
This is to understand that we cannot read actors' interests off their status
(Bevir, 2011) and that the business strategies of fish farming companies are
not 'so self-evident that … the researcher can easily identify them without
needing to conduct original empirical research' (Jullien and Smith, 2014: 5).
My intention when analysing the data therefore has not only been to produce
results on how sustainability choices are institutionalised and how actors are
mobilising around the politics of interdependence during their institutional
and political work, but also to document and render visible actors' social representations of these interdependencies, including their experiences of their
contradictions, conflicts and tensions. As the political scientist Robert Cox
said in his interview with Philippe Schouten:

> Collingwood spoke about the 'inside' as well as the 'outside' of historical
> events. When the positivist looks at what happens by classifying and col-
> lecting events and drawing inferences from them, he sees the outside;
> Collingwood's emphasis on the inside of events was to understand the
> meaning of things in terms of the thought-processes of the people who

were acting, and their understanding of the structure of relationships within which they lived. To understand history in those terms is what gives meaning to events.

(Cox speaking to Schouten: Schouten, 2009: 3)

The voices of the various stakeholders concerned serve to demonstrate that there is no single essential aquaculture, nor a universal and inevitable relationship between aquaculture and sustainability. Further, they also reveal the emotions and antagonistic political struggle taking place within governance (Mouffe, 2000). This includes making visible lines of division around which conflicts take place. I have given voice to these different representations by choosing interview citations which are representative of more general arguments, whilst respecting the anonymity of interviewees. In the presentation of findings in Chapters 4, 5 and 6, the reader will therefore be able to read actors' own words on the issue as they were experiencing it at the time of the interview.

Organisation of the book

The book is organised into two main parts. Part I sets out the analytical and conceptual framework of sustainability interdependence. In Chapter 2, I revisit both a long-standing and more recent literature on sustainability, distilling from it four 'ideal-type' narratives of sustainability which I present in turn. These are: (i) sustainability as depoliticised ecology–economy interdependencies; (ii) sustainability as politicised ecology–economy–society interdependencies; (iii) sustainability as ecosystem approaches problematising dynamic ecology–society interdependencies; and (iv) sustainability as degrowth re-ordering ecology–society–economy hierarchies.

Each narrative views and theorises interdependencies between ecology, economy and society in differing ways. I emphasise how the materialisation of 'sustainability' on the international policy agenda in the 1980s was underpinned by research which both constructed relations between nature and society as 'interdependencies' and debated ways in which these interdependencies could most effectively be governed. I highlight that, whereas different sustainability literatures hold in common a reappraisal of nature–society relations as interdependencies, they nonetheless differ on two main counts which will be the focus of my attention. First, on the content of key nature–society interdependencies considered to be at stake, including their hierarchy; and second, on the respective theories of politics underlying their analysis and the consequent character of institutional reform deemed necessary to govern these interdependencies within specific policy directions.

In Chapter 3 I set out how I have actually studied European fish farming aquaculture. A particular aim of the chapter is to recompose 'sustainability' analytically through integrating it into a wider theory of political change and a dynamic institutionalist framework of sustainability interdependence. I set out why it is important to develop a new framework for grasping sustainability, drawing from the public policy analysis literature highlighting the importance of policy–polity dialectics in political change and industry-governance. I expand the parameters of my analysis to include three interdependencies distributing power across the political system. These concern: (i) territory; (ii) regulation; (iii) knowledge. Finally, having described the contents and politics of these political interdependencies, I finish by explaining the conceptual analytical tools which I will mobilise in the empirical research. Ultimately, this is to reframe the research object as 'sustainability interdependence', through defining sustainability as an institution (public action principle) and integrating it within a wider theory of political change.

Part II of the book presents the empirical material demonstrating the implementation of sustainability in practice and the politics of 'sustainability interdependence' at stake. It is organised into three chapters, each addressing a problem-specific case study as described above. In Chapter 4, I examine political work on fish farm/environment interactions in Scottish salmon, Aquitaine trout and Greek seabass and seabream compared. In Chapter 5, I study political work on access to fish farm sites. In Chapter 6, I assess sustainability as a food governance problem. Through applying my analytical grid, the overall aim of these empirical chapters is to compare how actors in different territories and sectors, and in response to different sets of problems, have defined sustainability and rendered it governable over time.

I finish the book by drawing some general conclusions on the application of my analytical framework. I reconstitute my findings to summarise different country approaches to governance, drawing out overall similarities and differences between Scotland, Aquitaine and Greece. Finally, I conclude on how a 'sustainability interdependence' approach can open up new ways of debating aquaculture's place in the economy and society.

Overall

In summary, this book seeks to make two main contributions: first, to develop and apply a new analytical framework of 'sustainability interdependence' to grasp the politics of the implementation of sustainability; and second, to apply this framework to European fish farming aquaculture in different sectors and territories compared to generate original findings on actual governing practices addressing political controversies in aquaculture, including the political consequences of the policy directions chosen. Of course, the empirical material presented in this book is not exhaustive since it is limited both in time and in space. The account offered here on the governing of sustainability in European fish farming aquaculture is an account which emerges from an analysis of the material collected (Pinson and Sala Pala, 2007). It is an account which has been generated by the asking and answering of certain questions and the use of certain methods – and it invites other accounts (Yanow, 2007).

Bearing this in mind, the scientific posture I have adopted throughout is not one which is attempting to make a plea either for or against fish farming aquaculture as an industry. A strong plea is made, however, for an analytical discussion of the governing of this industry – as distinct from a normative and polemic one. Such a discussion can, I believe, be informed by new findings on the diversities of sustainability interdependence currently being institutionalised in Europe, including their points of contradiction and how actors experience them. The book therefore adopts a political and engaged positioning in this broader sense: through generating novel data on this emergent and highly contested industry, it seeks to open the door for new ways of debating European fish farming aquaculture.

The book demonstrates (i) how these debates ultimately came to shape the way in which sustainability problems in fish farming are rendered governable and, in turn, (ii) how the construction of sustainability as a governing problem for aquaculture is affecting broader processes of political and societal transformation. This critical distance in focus seeks to avoid any normative positioning of the book in relation to this industry (Eckersley, 2004).

Notes

1 Biotechnological innovation has produced novel forms of fish breeding, farming and ranching; for example, in Egypt, France, Norway, Scotland, Chile, Greece, Vietnam, China, Southern Australia, Spain, Africa, Canada.

2 For example, an important scholarship exists on fish farming aquaculture's environmental impacts and potential ecological and technical solutions (Belias *et al.*, 2003; Folke and Kautsky, 1989; Imsland *et al.*, 2014; Murry and Hall, 2014; Skiftesvik *et al.*, 2013); on aquaculture and ecosystem approaches to production and analysis (Costa-Pierce, 2010; Ferreira *et al.*, 2014; Grant, 2010; Nunes *et al.*, 2011); on technical and genetic innovation (Dunham, 2014; Shainee *et al.*, 2013); on fish nutrition (Barroso *et al.*, 2014; Fuertes *et al.*, 2013): on consumer behaviour and the marketing of farmed fish products (Ahmed *et al.*, 2009; Charles and Paquotte, 1999; Young *et al.*, 1999); and on the potential for smart regulation, integrated policy tools and sustainable policy to manage aquacultural practices (Boyd and Schmittou, 1999; Bunting, 2013; Howlett and Rayner, 2004; Kaiser and Stead, 2002). Contributions from the social sciences have additionally focused, for example, on tensions and synergies within aquaculture and fisheries interactions (Swanson, 2015; Wiber *et al.*, 2012); on comparative European legal regimes and planning laws and policies in specific territories (Glenn and White, 2007; Peel and Lloyd, 2008); and on the negative regulatory effects on US fish farming's global competitiveness (Engle and Stone, 2013).

3 I am paraphrasing here from Kauppi *et al.*, 2016.

4 In Chapter 2, I separate out debates on sustainability into four narratives on nature–society interdependencies. These are: (i) sustainability as depoliticised ecology–economy interdependencies; (ii) sustainability as politicised ecology–economy–society interdependencies; (iii) sustainability as ecosystem approaches problematising dynamic ecology–society interdependencies; and (iv) sustainability as de-growth re-ordering ecology–society–economy hierarchies.

5 This is also the case in other parts of the world, for example in Canadian fish farming aquaculture where demonstrating sustainability is 'the single most important production criterion' (Castle and Culver, 2008: 93).

6 This relative consensus has been stabilised at a global scale with the entering into force of the 'Illegal, Unreported and Unregulated' (IUU) fishing convention, although of course, this is not to suggest that all parties respect these rules.

7 Although in some cases, fish farming does contribute to restoration. For example, in Aquitaine trout farming, small fish farming companies can and do farm fish with the purpose of selling them to recreational fisheries associations to re-stock rivers.

8 www.dailymotion.com/video/x1u00l4_australia-with-simon-reeve-episode-1_travel.

9 www.dailymotion.com/video/x1u00l4_australia-with-simon-reeve-episode-1_travel.

10 See Chapter 2 for a full discussion of this literature.

11 In this book I am interested in interdependencies between different forms of institutionalisations, i.e. between institutions (rules and norms) and the actors who work in them. I explain this more fully in the later sections of this introductory chapter as well as in the chapters which immediately follow.

12 Although I do reconstitute territory-specific trajectories in the general conclusions in Chapter 7.

13 Spain, with 21 per cent of the total EU production in volume, is the largest aquaculture producer in the EU, followed by France (18 per cent), United Kingdom (14 per cent), Italy (13 per cent) and Greece (11 per cent). These five countries account for more than 75 per cent of the total EU aquaculture production in weight. In terms of value, the United Kingdom is the largest EU producer with 20 per cent of the total EU aquaculture, followed by France (19 per cent), Greece (15 per cent), Spain (12 per cent), and Italy (10 per cent). These five countries are also responsible for more than ¾ of all the EU aquaculture value.

(Marine Scotland, 2014: 11)

14 See Carter and Cazals (2014) for an account of why this is the case.

15 E.g. the Global Alliance Against Industrial Action (GAAIA; www.gaaia.org/) or Amper-TOS (www.peche-et-riviere.org/protection-riviere-tos.htm). See, for example, the recent report published by GAAIA on chemical treatment use in Scottish salmon production.

References

Agúndez, J., Raux, P., Girard, S., Mongruel, R. 2013. 'Technological adaptation to harmful algal blooms: socioeconomic consequences for the shellfish farming sector in Bourgneuf Bay (France)', *Aquaculture Economics & Management*, 17(4): 341–359.

Ahmed, N., Lecouffe, C., Allison, E., Muir, J. 2009. 'The sustainable livelihoods approach to the development of freshwater prawn marketing systems in southwest Bangladesh', *Aquaculture Economics & Management*, 13(3): 246–269.

Barroso, F., de Haro, C. Sánchez-Muros, M-J., Venegas, E., Martínez-Sánchez, A., Pérez-Bañón, C. 2014. 'The potential of various insect species for use as food for fish', *Aquaculture*, 422/423: 193–201.

Bartley, T. 2007. 'Institutional emergence in an era of globalization: the rise of transnational private regulation of labour and environmental conditions', *American Journal of Sociology*, 113(2): 297–351.

Belias, C., Bikas, V., Dassenakis, M., Scoullos, M. 2003. 'Environmental impacts of coastal aquaculture in eastern Mediterranean bays: the case of Astakos Gulf, Greece', *Environmental Science and Pollution Research*, 10(5): 287–295.

Benson, M., Craig, R. 2014. 'The end of sustainability', *Society & Natural Resources*, 27(7): 777–782.

Bevir, M. 2011. 'Public administration as storytelling', *Public Administration*, 89(1): 183–195.

Boyd, C., Schmittou, H. 1999. 'Achievement of sustainable aquaculture through environmental management', *Aquaculture Economics & Management*, 3(1): 59–69.

Bunting, S. 2013. *Principles of Sustainable Aquaculture*, Abingdon: Routledge.

Bureau, D. 2006. 'Rendered products in fish aquaculture feeds'. In Meeker, D. (ed.) *Essential Rendering: All about the Animal By-products Industry*, Arlington: The National Renderers Association; The Fats and Proteins Research Foundation; The Animal Protein Producers Industry, pp. 180–194.

Bush, S., Duijf, M. 2011. 'Searching for (un)sustainability in pangasius aquaculture: a political economy of quality in European retail', *Geoforum*, 42: 185–196.

Bush, S., Tri Khiem, N., Xuan Sinh, L. 2009. 'Governing the environmental and social dimensions of pangasius production in Vietnam: a review', *Aquaculture Economics & Management*, 13(4): 271–293.

Carter, C. 2012. 'Integrating sustainable development in the European government of industry: sea fisheries and aquaculture compared'. In Shuibhne, N. and Gormley, L. (eds) *From Single Market to Economic Union*, Oxford: Oxford University Press.

Carter, C. 2015. 'Who governs Europe? Public versus private regulation of sustainability of fish feeds', *Journal of European Integration*, 37(3): 335–352.

Carter, C., Cazals, C. 2014. 'The EU's government of aquaculture: completeness unwanted'. In Jullien, B. and Smith, A. (eds) *The EU's Government of Industries: Markets, Institutions and Politics*, Abingdon: Routledge, pp. 84–114.

Carter, C., Cazals, C., Hildermeier, J., Michel, L., Villareal, A. 2014a. 'Sustainable development policy: "Competitiveness" in all but name'. In Jullien, B. and Smith, A. (eds) *The EU's Government of Industries: Markets, Institutions and Politics*, Abingdon: Routledge, pp. 165–189.

Carter, C., Ramírez, S., Smith, A. 2014b. 'Trade policy: all pervasive but to what end?'. In Jullien, B. and Smith, A. (eds) *The EU's Government of Industries: Markets, Institutions and Politics*, Abingdon: Routledge, pp. 216–240.

Castle, D., Culver, K. (eds) 2008. *Aquaculture, Innovation and Social Transformation*, Netherlands: Springer.

Charles, E., Paquotte, P. 1999. 'Product differentiation, labelling and quality approach: developments and stakes in the French shellfish market', *Aquaculture Economics & Management*, 3(2): 121–129.

Cleaver, F., Franks, T. 2005. 'How institutions elude design: river basin management and sustainable livelihoods'. Bradford Centre for International Development, Research Paper no. 12. Available at: http://core.kmi.open.ac.uk/display/5659.

Costa-Pierce, B. 2008. 'An ecosystem approach to marine aquaculture: a global review'. In Soto, D., Aguilar-Manjarrez, J. and Hishamunda, N. (eds) *Building an Ecosystem Approach to Aquaculture*, FAO/Universitat de les Illes Balears Experts Workshop. 7–11 May 2007, Palma de Mallorca, Spain. FAO Fisheries and Aquaculture Proceedings No. 14, Rome: Food and Agriculture Organisation (FAO), pp. 81–115.

Costa-Pierce, B. 2010. 'Sustainable ecological aquaculture systems: the need for a new social contract for aquaculture development', *Marine Technology Society Journal*, 44(3): 88–112.

Crawford, M. 2010. 'The importance of long chain omega-3 fatty acids for human health'. Paper presentation at IG-CCBSD conference 'Can a Growing Aquaculture Industry Continue to use Fishmeal and Fish Oil in Feeds and Remain Sustainable?', European Parliament, Brussels, 3 March.

Dunham, R. 2014. 'Introduction to genetics in aquaculture XI: the past, present and future of aquaculture genetics', *Aquaculture*, S1/S2: 420–421.

Eckersley, R. 2004. *The Green State: Rethinking Democracy and Sovereignty*, Cambridge MA: MIT Press.

Engle, C., Stone, N. 2013. 'Competitiveness of U.S. aquaculture within the current U.S. regulatory framework', *Aquaculture Economics & Management*, 17(3): 251–280.

Ernst & Young. 2008. *Etude des performances économiques et de la compétitivité de l'aquaculture de l'Union européenne*, Report for DG MARE, December.

European Commission of the European Communities. 2009a. *Building a Sustainable Future for Aquaculture*, Communication, Com(2009)162, Brussels, 8 April.

European Commission of the European Communities. 2009b. *Impact Assessment in Respect of Com(2009)162*, SEC(2009)453, Brussels.

Ferreira, J., Saurel, C., Lencart e Silva, J., Nunes, J., Vazquez, F. 2014. 'Modelling of interactions between inshore and offshore aquaculture', *Aquaculture*, 426/427: 154–164.

Folke, C., Kautsky, N. 1989. 'The role of ecosystems for a sustainable development of aquaculture', *AMBIO: A Journal of the Human Environment*, 18(4): 234–243.

François, P. (ed.) 2011. *Vie et mort des institutions marchandes*, Paris: Presses de Sciences Po.

Frickel, S., Moore, K. (eds) 2006. *The New Political Sociology of Science Institutions, Networks, and Power*, Madison, WI: The University of Wisconsin Press.

Fuertes, J., Celada, J., Carral, J., Sáez-Royuela, M., González-Rodríguez, A. 2013. 'Replacement of fish meal with poultry by-product meal in practical diets for juvenile crayfish (*Pacifastacus leniusculus* Dana, Astacidae) from the onset of exogenous feeding', *Aquaculture*, 404/405: 22–27.

Genieys, W., Smyrl, M. 2008. *Elites, Ideas and the Evolution of Public Policy*, London and New York: Palgrave.

Gibbs, M. 2009. 'Implementation barriers to establishing a sustainable coastal aquaculture sector', *Marine Policy*, 33: 83–89.

Glenn, H., White, H. 2007. 'Legal traditions, environmental awareness, and a modern industry: comparative legal analysis and marine aquaculture', *Ocean Development and International Law*, 38: 71–99.

Grant, A.N., Treasurer, J.W. 1993. 'The effects of fallowing on caligid infestations on farmed Atlantic salmon (Salmo salar L.) in Scotland', in Boxshall, G.A. and Defaye, D. (eds) *Pathogens of Wild and Farmed Fish: Sea Lice*, New York: Ellis Horwood Ltd., pp. 255–260.

Grant, J. 2010. 'Coastal communities, participatory research, and far-field effects of aquaculture', *Aquaculture Environment Interactions*, 1: 85–93.

Hay, C., 2010. 'Introduction: political science in an age of acknowledged interdependence'. In Hay, C. (ed.) *New Directions in Political Science: Responding to the Challenges of an Interdependent World*, Basingstoke: Palgrave, pp. 1–24.

Healey, P. 2004. 'The treatment of space and place in the new strategic spatial planning in Europe', *International Journal of Urban and Regional Research*, 28(1): 45–67.

Hites, R., Foran, J., Carpenter, D., Hamilton, M., Knuth, B., Schwager, S. 2004. 'Global assessment of organic contaminants in farmed salmon', *Science*, 303: 226–229.

Hopwood, B., Mellor, M., O'Brien, G. 2005. 'Sustainable development: mapping different approaches', *Sustainable Development*, 13: 38–52.

Howlett, M., Rayner, J. 2004. '(Not so) "smart regulation"? Canadian shellfish aquaculture policy and the evolution of instrument choice for industrial development', *Marine Policy*, 28: 171–184.

Huising, R., Silbey, S. 2011. 'Governing the gap: forging safe science through relational regulation', *Regulation & Governance*, 5: 14–42.

Huntington, T.C. 2004. *Feeding the Fish: Sustainable Fish Feed and Scottish Aquaculture*, Report to the Joint Marine Programme Scottish Wildlife Trust and WWF Scotland and RSPB Scotland.

Imsland, A., Reynolds, P., Eliassen, G., Hangstad, T., Foss, A., Vikingstad, E., Elvegård, T. 2014. 'The use of lumpfish (Cyclopterus lumpus L.) to control sea lice (Lepeophtheirus salmonis Krøyer) infestations in intensively farmed Atlantic salmon (Salmo salar L.)', *Aquaculture*, 424/425: 18–23.

Jeffery, C., Wincott, D. 2010. 'The challenge of territorial politics: beyond methodological nationalism'. In Hay, C. (ed.) *New Directions in Political Science: Responding to the Challenges of an Interdependent World*, Basingstoke: Palgrave, pp. 167–188.

Jobert, B., Muller, M. 1987. *L'Etat en action*, Paris: PUF.

Jordan, A., Adelle, C. (eds). 2013. *Environmental Policy in the EU: Actors, Institutions and Processes*, 3rd edition, Abingdon: Routledge.

Jullien, B., Smith, A. (eds). 2014. *The EU's Government of Industries: Markets, Institutions and Politics*, Abingdon: Routledge.

Kaiser, M., Stead, S. 2002. 'Uncertainties and values in European aquaculture: communication, management and policy issues in times of changing public perceptions', *Aquaculture International*, 10: 469–490.

Kauppi, N. 2010. 'The political ontology of European integration', *Comparative European Politics*, 8(1): 19–36.

Kauppi, N., Palonen, K., Wiesner, C. 2016. 'Controversy in the garden of concepts: rethinking the "politicisation" of the EU'. Joint Working Paper Series of Mainz Papers on International and European Politics (MPIEP) No. 11 and Jean Monnet Centre of Excellence 'EU in Global Dialogue' (CEDI) Working Paper Series No. 3. Mainz: Chair of International Relations, Johannes Gutenberg University.

Lafferty, W. (ed.). 2004. *Governance for Sustainable Development*, Cheltenham: Edward Elgar Publishing.

Lafferty, W., Hovden, E. 2003. 'Environmental policy integration: towards an analytical framework', *Environmental Politics*, 12(3): 1–22.

Lange, P., Driessen, P., Sauer, A., Bornemann, B., Burger, P. 2013. 'Governing towards sustainability: conceptualizing modes of governance', *Journal of Environmental Policy & Planning*, 15(3): 403–425.

Lawrence, T., Suddaby, R. 2006. 'Institutions and institutional work'. In Hardy, C., Lawrence, T. and Nord, W. (eds) *Sage Handbook of Organization Studies*, 2nd edition, London: Sage, pp. 215–254.

Marine Scotland. 2014. *MGSA Science & Research Working Group Aquaculture Science & Research Strategy*, produced on behalf of the Scottish Government Ministerial Group for Sustainable Aquaculture (MGSA) May 2014.

Merand, F. 2008. 'Les institutionnalistes (américains) devraient-ils lire les sociologues (français)?', *Politique européenne*, 2(25): 23–51.

Mouffe, C. 2000. *Deliberative Democracy or Agonistic Pluralism*. Wien: (Reihe Politikwissenschaft/Institut für höhere Studien, Abt. Politikwissenschaft 72). Available at: http://nbnresolving.de/urn:nbn:de:0168-ssoar-246548.

Muller, P., 2010. 'Les changements d'échelles des politiques agricoles: introduction'. In Hervieu, B., Mayer, N., Muller, P., Purseigle, F. and Rémy, J. (eds) *Les Mondes Agricoles en Politique*, Paris: Presses de Sciences Po, pp. 339–350.

Muller, P. 2015. *La société de l'efficacité globale: comment les sociétés se pensent et agissent sur elles-mêmes*, Paris: PUF.

Murry, A., Hall, M. 2014. 'Treatment rates for sea lice of Scottish inshore marine salmon farms depend on local (sea loch) farmed salmon biomass and oceanography', *Aquaculture Environment Interactions*, 5(2): 117–125.

Naylor, R.L., Goldburg, R.J., Primavera, J.H., Kautsky, N., Beveridge, M.C.M., Clay, J., Folke, C., Lubchenco, J., Mooney, H., Troell, M. 2000. 'Effect of aquaculture on world fish supplies', *Nature*, 405 (6790): 1017–1024.

Noakes, D., Fang, F., Hipel, K., Kilgour, D. 2003. 'An examination of the salmon aquaculture conflict in British Columbia using the graph model for conflict resolution', *Fisheries Management and Ecology*, 10: 123–137.

Nunes, J.P., Ferreira, J.G., Bricker, S.B., O'Loan, B., Dabrowski, T., Dallaghan, B., Hawkins, A.J.S., O'Connor, B., O'Carroll, T. 2011. 'Towards an ecosystem approach to aquaculture: assessment of sustainable shellfish cultivation at different scales of space, time and complexity', *Aquaculture*, 315: 369–383.

Peel, D., Lloyd, M. 2008. 'Governance and planning policy in the marine environment: regulating aquaculture in Scotland', *The Geographical Journal*, 174(4): 361–373.

Pinson, G., Sala Pala, V. 2007. 'Peut-on vraiment se passer de l'entretien en sociologie de l'action publique?', *Revue française de science politique*, 5(57): 555–597.

Porter, G. 2005. *Protecting Wild Atlantic Salmon from Impacts of Salmon Aquaculture: A Country-by-Country Progress Report*. Second report. Available at: www.asf.ca/Communications/2005/05/impacts2005.pdf.

Rana, K. 2007. *Regional Review on Aquaculture Development 6: Western-European Region-2005*, FAO Fisheries Circular No. 1017/6, FIMA/C1017/6, Rome: Food and Agriculture Organisation (FAO).

Russell, M., Robinson, C., Walsham, P., Webster, L., Moffat, C., 2011. 'Persistent organic pollutants and trace metals in sediments close to Scottish marine fish farms', *Aquaculture*, 319: 262–271.

Schneiberg, M. 2007. 'What's on the path? Path dependence, organizational diversity and the problem of institutional change in the US economy, 1900–1950', *Socio-Economic Review*, 5(1): 47–80.

Schouten, P. 2009. 'Theory Talk #37: Robert Cox on world orders, historical change, and the purpose of theory in international relations', *Theory Talks*. Available at: www.theorytalks.org/2010/03/theory-talk-37.html.

Scottish Government. 2010. *Scottish Fish Farm Production Survey 2010*, Edinburgh: Scottish Government. Available at: www.gov.scot/Topics/marine/Publications/stats/FishFarmProductionSurveys/OlderSurveys.

Shainee, M., Ellingsen, H., Leira, B., Fredheim, A. 2013. 'Design theory in offshore fish cage designing', *Aquaculture*, 392/395: 134–141.

Skiftesvik, A., Bjelland, R., Durif, C., Johansen, I., Browman, H. 2013. 'Delousing of Atlantic salmon (Salmo salar) by cultured vs. wild ballan wrasse (Labrus bergylta)', *Aquaculture*, 402/403: 113–118.

Soto, D., Aguilar-Manjarrez, J., Brugère, C., Angel, D., Bailey, C., Black, K., Edwards, P., Costa-Pierce, B., Chopin, T., Deudero, S., Freeman, S., Hambrey, J., Hishamunda, N., Knowler, D., Silvert, W., Marba, N., Mathe, S., Norambuena, R., Simard, F., Tett, P., Troell, M., Wainberg, A. 2008. 'Applying an ecosystem-based approach to aquaculture: principles, scales, and some management measures'. In Soto, D., Aguilar-Manjarrez, J. and Hishamunda N. (eds) *Building an Ecosystem*

Approach to Aquaculture. FAO/Universitat de les Illes Balears Experts Workshop. 7–11 May 2007, Palma de Mallorca, Spain. FAO Fisheries and Aquaculture Proceedings No. 14, Rome: Food and Agriculture Organisation (FAO), pp. 15–35.

Swanson, H. 2015. 'Shadow ecologies of conservation: co-production of salmon landscapes in Hokkaido, Japan, and southern Chile', *Geoforum*, 61: 101–110.

Tacon, A.G.T., Metian, M., Turchini, G.M., de Silva, S.S. 2010. 'Responsible aquaculture and trophic level implications to global fish supply', *Reviews in Fisheries Science*, 18(1): 94–105.

Tiller, R., Brekken, T., Bailey, J. 2012. 'Norwegian aquaculture expansion and Integrated Coastal Zone Management (ICZM): simmering conflicts and competing claims', *Marine Policy*, 36: 1086–1095.

Wiber, M., Young, S., Wilson, L. 2012. 'Impact of aquaculture on commercial fisheries: fishermen's local ecological knowledge', *Human Ecology*, 40: 29–40.

Yanow, D. 2007. 'Interpretation in policy analysis: on methods and practice', *Critical Policy Studies*, 1(1): 110–122.

Young, J., Brugere, C., Muir, J. 1999. 'Green grow the fishes-oh? Environmental attributes in marketing aquaculture products', *Aquaculture Economics & Management*, 3(1): 7–17.

Young, J.C., Waylen, K.A., Sarkki, S., Albon, S., Bainbridge, I., Balian, E., Davidson, J., Edwards, D., Fairley, R., Margerison, C., McCracken, D., Owen, R., Quine, C., Stewart-Roper, C., Thompson, D., Tinch, R., Van den Hove, S., Watt., A. 2014. 'Improving the science–policy dialogue to meet the challenges of biodiversity conservation: having conversations rather than talking at one-another', *Biodiversity and Conservation* 23: 387–404.

Part I

Theorising a politics of sustainability interdependence

2 Sustainability narratives of nature–society interdependencies and the re-organisation of state power

Introduction

One of the central challenges facing society today is its response to ecological imperatives, including over ocean health, water quality and biodiversity (Rudd, 2014). However, what place does European fish farming aquaculture have within this societal response? As I set out in the introductory chapter, this book considers this question from the governing perspective. This is to assume that there is no self-evident relationship either between fish farming aquaculture and society or between fish farming aquaculture and nature, but that these relationships depend on the way in which the industry is governed. A first step therefore in the inquiry is to establish the analytical grid for grasping the governing politics of this industry. The first part of the book does precisely that. Drawing on arguments which highlight policy–polity tensions as critical for understanding governing choices, I make the case for updating and recomposing the concept of sustainability, integrating it into a wider political sociological theory of political change and reframing the research object as 'sustainability interdependence'.

Part I is organised into two chapters. In this first chapter, I revisit both long-standing and more recent arguments in an extensive literature on sustainability spanning several decades. As Jordan already argued in 2008, the literature on sustainability is 'truly vast' (Jordan, 2008) and has repeatedly shown that sustainability is a 'contested concept' (Giddings *et al.*, 2002: 187) with a 'multiplicity of meanings' (Haughton and Counsell, 2004: 72–73) in different settings (see also: Baker *et al.*, 1997; Dobson, 1996; Dryzek, 1997; Hajer 2005 [1995]; Jordan, 2008; Jordan and Lenschow, 2010; Pestre, 2011; Scoones, 2016; Springett, 2013;). This chapter does not attempt a comprehensive summary of this literature. Rather, through distilling from it four conceptual 'ideal-type' narratives of sustainability interdependencies, I underline that the emergence of 'sustainability' in the 1980s was accompanied by arguments which not only forced a reappraisal of relations between nature and society, re-casting them as 'interdependencies', but also encouraged debate on the appropriate role for government (and the state) governing the interdependencies identified.

Summarising how the different theoretical and conceptual approaches have apprehended sustainability's association, first with the notion of 'interdependence', and second with the reform of the nation state and its governing institutions, I explore each narrative in turn. These are: (i) sustainability as depoliticised ecology–economy interdependencies; (ii) sustainability as politicised ecology–economy–society interdependencies; (iii) sustainability as ecosystem approaches problematising dynamic ecology–society interdependencies; and (iv) sustainability as de-growth re-ordering ecology–society–economy hierarchies.

In this manner, I aim to shift the focus of inquiry from 'sustainability' to the interdependencies and their narratives.[1] I highlight that, whereas different sustainability literatures hold in common a reappraisal of nature–society relations cast as interdependencies, they nonetheless differ on two main counts which will be the focus of attention: first, on the contents of key nature–society interdependencies considered to be at stake, including their hierarchy; and second, on respective theories of politics underlying their analysis and the consequent character of institutional reform deemed necessary to govern these interdependencies within specific policy directions. Indeed, just 'how' the political change associated with sustainability would occur, 'how much' change was necessary, and 'which values' (Smith, 2016) should be upheld in the policy choices induced, are all questions which have driven (and continue to drive) conflicts within sustainability theories and politics:

> Environmental politics is only partially a matter of whether or not to act, it has increasingly become a conflict of interpretation in which a complex set of actors can be seen to participate in a debate.
>
> (Hajer, 2005 [1995]: 15)

Overall, the chapter plays a key role in contributing to the line of argumentation to be advanced in this book. First, through clarifying the various meanings of sustainability and updating them around narratives of interdependencies, I move away from syntheses of 'sustainability' which categorise it in terms of 'strong' versus 'weak' or 'light green' versus 'dark green'. These metaphorical approaches to categorisation are often confusing and create problems when seeking to manage analyses of empirical material. Instead, I have chosen to organise each narrative around its core components which provide key points of difference and which can be mobilised to categorise my own empirical material (see Chapter 3). Second, through disentangling a contested concept into different narratives of interdependence, I can more easily isolate the finer details of difference in governing practices within European fish farming aquaculture. This will not only help me to make clear comparisons between the different sectors and territories studied, but also to grasp contingency in sustainability governing through raising awareness of possible alternatives. I return to these points in the final section of Chapter 3.

Sustainability as depoliticised ecology–economy interdependencies

As is well-documented, the 1970s and 1980s were marked by what Hajer has termed an 'ecological turn' in global politics (Hajer 2005 [1995]). During this period, a series of environmental conditions, for example water, air and soil pollution, were defined as public problems (Carson, 1962, Meadows *et al.*, 1972), mobilised by a growing grass roots environmental movement (Faucher, 1999) and debated on an international scale (e.g. in the UN conference on the Human Environment in Stockholm, 1972). During this intense activity on the ecological question, both in science and in politics, different phenomena hitherto held in relation to one another began to be re-problematised as 'interdependent'. Yet, as will be shown repeatedly, whereas analytically in this book I will treat 'interdependence' as an object to be analysed, many works on sustainability have instead treated interdependencies in a prescriptive way.

From the 1960s to the1980s, a central such interdependence which began to be recognised was that between ecology and economy. Whereas 'economists have known for a century that human activities could alter the environment' (Brooks, 1992: 402), during this period it became increasingly accepted amongst different communities of actors that the reverse was also possible, namely that the environment could set limits on the economy. According to Gómez-Baggethun and Naredo (2015), that natural resources and biophysical processes could place their own limits on economic growth was an understanding which had first emerged in the nineteenth-century French physiocratic movement, but it was disregarded in the neoclassical economic thinking which came to dominate economics at the beginning of the twentieth century. However, during the 1960s and 1970s, and in their critique of mainstream economic thinking, scholars such as Kapp (1961), Boulding (1966) Georgescu-Roegen (1971) and Daly (1973) began to challenge the viability of the perpetual growth model inherent in neo-classical economics and to re-theorise the claim that the environment set its own limits to growth (this work would provide the foundation for the sub-discipline of ecological economics, Asara *et al.*, 2015).[2]

Against this global background of a burgeoning environmental politics, the publication of the 1987 Report of the World Commission on Environment and Development (WCED) *Our Common Future* (known as the Brundtland Report) has been considered critical as the 'linchpin in the creation of the new consensus' around the political project of 'sustainable development' responding to global ecological crises emerging at that time (Hajer 2005[1995]: 1).[3]

Although not adopting the more radical 'limit to growth' policy advocated by some ecological economists,[4] Brundtland (as the Report is commonly referred to) nonetheless did endorse the 'shift from economic activities affecting the environment (the externality model) to the environment affecting economic activities (the carrying capacity model)' as theorised by these scholars (Brooks, 1992: 402). Brundtland not only argued that economic development had incurred 'environmental stress', whereby thresholds for eco-system integrity were being reached, but also that this (potentially irrevers-ible) ecological damage was in turn placing limits on future economic development (WCED 1987: 44, 47). In this way, Brundtland both 'rendered visible' (Jouzel and Dedieu, 2013) a new interdependence between economy and ecology and formalised it on a global scale: economy and ecology rela-tionships were no longer considered one way or dependent (Hay, 2010), but 'interactive' (Brooks, 1992: 402).

The theorisation of interdependencies between ecology and economy did not only take place within economics, however. Rather, the emergence of 'ecological modernisation' theory within political science and environ-mental sociology in the 1980s was to prove critical, not only for influencing Brundtland (Hajer, 2005 [1995]), but for influencing policy and industrial practice for years to come (Carter *et al.*, 2014; Dryzek, 1997; Giddings *et al.*, 2002; Gómez-Baggethun and Naredo, 2015; Jordan and Adelle, 2013; Toke, 2011; Weale and Williams, 1992). Founded in the 1980s,[5] ecological modernisation has evolved as a theory over time, responding especially to critique of its minimal reformist agenda, its theory of causal mechanisms of change and its failure to address questions of social justice. This critique has led in turn to the development of two distinct strands within this approach: what Hajer has referred to as a 'techno-corporatist' versus a 'reflexive' dis-course of ecological modernisation (2005 [1995]: 164–165); what Christoff has described as a 'weak' versus a 'strong' form of ecological modernisation (1996); and what Horlings and Marsden have described as two distinct forms of ecologically modernised green economies – the bio-economy versus the eco-economy (2011, 2014). In the rest of this sub-section, I discuss the techno-corporatist form, returning to the reflexive form in the next sub-section.

In contrast with early (and later) intentions of some ecological economists, the central philosophy of early ecological modernisation theorists was to *depo-liticise* any potential contradictions inherent within ecology–economy interde-pendencies. Techno-corporatist versions of this theory contended that ecological degradation could be *fixed* within the current system of liberal capitalism, assuming some reform of its main institutions (Buttel, 2000: 59). Scholars hypothesised that technological innovation coupled with market efficiency would provide central mechanisms for off-setting any limits imposed by nature (Brooks, 1992). Technological innovation would be the means through which industrial actors could continue to grow their business sustainably and manage any potential contradictions at the heart of the ecol-ogy–economy interdependence. Theoretically, technological innovation was expected to act as an independent causal process, automatically assuring 'eco-logical switchover', defined as the ecological modernisation of production and consumption (Spaargaren 2000, 48). This would decouple growth from environmental degradation and render it 'possible within the limits set by nature' (Langhelle, 2000: 310).[6]

This thesis was further nourished by an efficiency theory of the market place. Adjustments in markets to facilitate the uptake and eventual spread of new technology were initially considered to result from its 'invisible hand', coordinating the interests of various groups of actors, including consumers (constructed as passive) (Toke, 2011). This neoclassical economics view of markets was buttressed by evolutionary visions of the Schumpeterian innovation cycle consisting of invention, innovation and diffusion. The outcome would be 'win–win' across the interdependence: ecologically 'clean technology' solutions would be achieved through competition and profitability.

These evolutionary underpinnings become important in regard to how early ecological modernisation theorists subsequently began to view the role of governing mechanisms for industrial change through technological innovation. Extending their efficiency theory of markets to politics, a reduced role for government was anticipated (Grabosky, 1995). Ecological modernisation

> has therefore challenged the traditional role for states, for there is a perceived need for leverage, rather than try[ing] to micro-manage, the potential for environmental innovation within market economies.
>
> (Williamson and Lynch-Wood, 2012: 941)

This leverage would come in the form of new 'managerial techniques' (such as Environmental Impact Assessment) or technical economic tools (eco-taxes, pollution trading permits) to encourage switchover (Hopwood *et al.*, 2005: 43). Otherwise, the governance of the innovation cycle was understood to be brought about through self-regulatory initiatives of large multinational firms responding to a variety of legal, economic, organisational and social pressures to 'go green' (Huber, 2000) and entering into elite partnerships with governments and science (Christoff, 1996). Self-regulatory initiatives included the designing of 'environmental management systems' by individual companies and global leaders aimed at spreading technological innovation throughout their group or supply chain. These kinds of public/private governing mechanisms were identified and applauded by theorists as providing key examples of how ecological modernisation worked (Huber, 2000). Much later on, the role for government in these processes has been both summarised and clarified by Jänicke and Lindemann (2010). Continuing with the claim that technological innovation can bring about environmental improvement, the focus is on how to develop 'innovation-orientated environmental policy' (Jänicke and Lindemann 2010: 127).

In an article which does precisely that, Jänicke and Lindemann, set out the role for government as facilitating the 'co-evolution of socio-technological configurations' and the design of smart regulation to 'maximise the effectiveness of technological development' (2010: 129). Management tools such as 'technological forcing' are applauded whereby radical, as distinct from incremental, innovation with a high level of market penetration is expected to have greater environmental success (Jänicke and Lindemann 2010: 133). In these approaches, governmental challenges become ones of inter- and intra-policy coordination in order to ensure that technological innovation can meet its green target in different localities (Greenwood, 2015).

In summary, this first sustainability narrative depoliticises ecology–economy interdependencies wherein technological innovation is a panacea for environmental degradation. Reform of the state is necessary to reduce its role from one of 'provider' to 'enabler' (Davidson, 2012: 37), developing new governing institutions in the form of market instruments as economic incentives to innovate.

Sustainability as politicised ecology–economy–society interdependencies

There have been many conceptual challenges to sustainability as depoliticised ecology–economy interdependencies and many critiques of its empirical applications (York and Rosa, 2003). A first strong counter-narrative can be found in collective works self-positioning themselves as being within either a strong version of ecological modernisation, the sustainable development literature, or environmental governance.[7] This second narrative therefore emerges from disparate scholarly work, rather than from within the elaboration of a single theory – 'there is no such thing as … sustainable development "ism"' (Hopwood *et al.*, 2005: 47) – although it carries a strong Keynesian undercurrent. Taken as a whole, these works part company from sustainability as depoliticised ecology–economy interdependencies in two important ways. First, they redefine the contents of the interdependencies at stake in sustainability politics. Second, they seek to replace evolutionary and efficiency theories of sustainability politics with distributional ones, problematising both agency and societal democracy (Leroy and Van Tatenhove, 2000; Giddings *et al.*, 2002; Hopwood *et al.*, 2005).

This body of work first extends the parameters of 'interdependence' associated with sustainability to include social welfare and justice. In his article drawing a conceptual line between ecological modernisation and sustainable development, Langhelle argued that social justice is '*the* primary development goal of sustainable development' (Langhelle, 2000: 307). This is echoed by Hopwood *et al.* (2005): 'Social justice today and in the future is a crucial component of the concept of sustainable development.' In these analyses of sustainability, authors insist on the importance of considering economic, ecological and social relationships as interdependent. Any separation of these components, they argue, results in the undermining of human well-being at the heart of sustainability whereby 'wider social issues often fall off the sustainable development agenda' (Giddings *et al.*, 2002: 189–190). Taking inspiration from Brundtland's definition of sustainable development as meeting 'the needs of the present without compromising the ability of future generations to meet their needs' (WCED, 1987: 43), they consider equity to be a central guiding principle of sustainability.[8] Analysis of sustainability around the ecology–economy–society interdependence has therefore placed attention on questions of poverty and inequality, establishing connections between environmental degradation, human injustices and discrimination (Becker *et al.*, 1999; Deldrève and Candau, 2014; Hopwood *et al.*, 2005: 39; MacGregor, 2004).

These works not only expanded the contents of key interdependencies at stake in sustainability politics, they also sought to *politicise* potential contradictions between them:

> If sustainable development implies more than an efficiency-oriented approach to the environment, it is no longer necessarily the case that sustainable development represents a 'win–win' solution.
>
> (Langhelle, 2000: 316)

Acknowledging a potential 'hierarchy of priorities and weighing of different concerns inherent in the concept of sustainable development' (Langhelle, 2000: 318), the politics of sustainability was consequently defined in terms of choices:

> The vision is based on a construction of sustainable development that is problematic: not a discourse of environment and conservation and growing 'eco-cracy', nor one of eco-modernization and 'green business', but one of social crisis and human agency.
>
> (Springett, 2013: 79)

The politicisation of ecology–economy–society interdependencies carried with it a transformationist institutional and governing agenda in which we can identify two main components: first, a strong policy role for government with an emphasis on policy integration; and second, the need for democratic change towards participatory, deliberative and/or eco-democracy to open up new political spaces in which competing claims could be debated.

Hence, rather than a reduced role for government working on the market, this politicised vision of sustainability advocated an interventionist public policy working on sectors and production practices:

> Thus, depending on the principle of distribution, an environmental policy based on the paradigm of sustainable development may be a much more demanding and ambitious one than ecological modernization.
>
> (Langhelle, 2000: 317)

A central change anticipated was to move from a sectoral to an 'integrated' approach to regulation (Eckersley, 1995: 8; Giddings *et al.*, 2002: 189–190). This would not only require a change in the contents of sectoral policy, through its taking on board environmental questions, but also a reform of government administrations. Scholars argued that

> 'it is semantically inconsistent to conceive of sustainable development without successful EPI [environmental policy integration]' which is 'a central element of the concept of "sustainable development"'.
>
> (Lafferty and Hovden, 2003: 2, 1)

The notion of policy integration had been addressed in Brundtland, as well as the need to re-organise the way in which administrations handled policy affairs:

> Society has failed to give responsibility for preventing environmental damage to the 'sectoral' ministries and agencies whose policies cause it.
>
> (WCED, 1987 50)

Brundtland argued that most administration had reduced environmental departments' role to redressing environmental effects caused by others (WCED 1987: 25). The authors therefore called for a '[n]ew approach ... [one] that integrates production with resource conservation and enhancement' (WCED 1987: 50). This argument was repeated throughout, whereby mandates of sectoral departments had to be expanded to include environmental responsibilities (WCED 1987: 26, 50–51, 71).

Taking inspiration from Brundtland, yet criticising what they interpreted as its limited application to economic policy, scholars argued that 'real' sustainability would only come about through the application of environmental policy integration to all policy areas. Inherent in this argument was a normative expectation that through policy integration, environmental protection (and specifically the protection of the environment's carrying capacity) would be strongly institutionalised as a societal, as distinct from a sectoral, commitment. These scholars identified, for example, international and EU commitments to environmental policy integration (e.g. in the Rio Declaration and Agenda 21; in EU Environmental Action Programmes; in Article 6 of the Treaty on European Union, 1993; and in the Cardiff process)[9] as strong examples of the kind of 'constitutional commitment' expected (Lafferty and Hovden, 2003: 4).

Institutional and organisational reforms were deemed necessary to implement policy integration, whereby changes in both instruments as well as administrative capacity for governance were promoted to ensure both sectoral (or vertical) and inter-sectoral (or transversal) integration. For example, Lafferty and Hovden list the following as examples of what they describe as effective instruments and policy tools of environmental policy integration. In the case of sectoral policy integration, these include *inter alia* Environmental Impact Assessments, Strategic Environmental Assessments, sectoral sustainability action plans and mapping of sectoral environmental challenges. In the case of inter-sectoral integration, these include long-term sustainable development strategies, the creation of a central authority within a government whose responsibility it is to monitor the implementation of a sustainable development strategy, time-tables and periodic reviews, and environmental impact assessments of policy proposals (Lafferty and Hovden, 2003: 13–15: see also Jordan and Lenschow, 2008, 2010; Lenschow, 2002). In this context, scholars argued for strong public policy action in respect of the environmental 'precautionary principle': such action would not only place the burden of proof on business to demonstrate that any economic development requested would not harm the environment, but would also adopt a default position against development in the absence of knowledge on its ecological effects (European Environment Agency, 2001: 13).

Second, scholarship espousing the sustainability as politicised ecological-economic-social interdependencies narrative has also advocated broader democratic change to realise sustainability. As argued by Van Tatenhove and Leroy (2003), environmental political movements of the 1960s and 1970s had coupled their critique of the way in which environmental problems had been governed hitherto with a more general anti-modernist critique of the nation state. Seeing participation of citizens in decision-making as a corrective to elitist governing and administrative logics of action (Barbier and Larrue, 2011: 68), the green politics movement argued that new forms of democracy (other than representative democracy) were required to give expression to social justice norms at the heart of sustainability politics (Dobson, 1998). Overall, arguments in favour of participation have been premised upon expectations that participation would (i) bring about multiple benefits to natural resource management (Blackstock and Richards, 2007); (ii) introduce environmentally-friendly social values into policy decisions (Barbier and Larrue, 2011: 68); (iii) enhance the quality of any decisions taken (Reed, 2008); and iv) build legitimacy and trust in policy outcomes (Holmes and Scoones, 2000: 3).

Implicit in many of these proposals for 'suitable governance for SD [sus-tainable development]' (Lange *et al.*, 2013: 404) is a normative expectation that both environmental policy integration and participatory democracy will take place in a context where 'protecting the environment' is already a shared societal ambition. This might exist in the form of an already settled sustain-ability social contract, whereby 'protecting the environment' has been brought within the wider social bargain underpinning any policy choice. This underlying positioning has been taken to its extreme in the work of Eckersley (2004) and her theorisation of the 'green state'. Considering that a new environmental social contract is possible within a transformed liberal capitalist state, Eckersley has argued, first, that ecologically informed counter develop-ments[10] have already transformed the 'anti-ecological dynamics' of the state (2004: 241), and second, that these changes require to be consolidated in a fully-fledged transformation to a 'green' state:

This virtuous relationship however cannot be deepened without a move from liberal democracy to ecological democracy.

(Eckersley, 2004: 241)

This transformation will come about through what she describes as tensions between the functional and constitutive elements of ecological modernisation (Eckersley, 2004) – and what we might describe here as contradictions between sustainability narratives. But this transformation will only be stabilised if it takes place within a green political project seeking a 'green constitution' (Eckersley, 2004: 245) with green constitutional rights (equivalent in law to social and economic human rights). These include, for example the right to participate in Environmental Impact Assessments. Consequently, in this green state, not only has environmental policy integration been extended from being an organisational and institutional norm into a constitutional one, but there has also been a radical institutionalisation of a societal commitment to nature equivalent to social welfare. This has been explicitly set out by Duit *et al.*, who argue that heuristically the 'environmental state' can be thought of as a theoretical abstraction in the same way as 'the welfare state' has been, referring both to its institutions and the wider polity in which those institutions are exercised (2016).

In summary, this sustainability narrative politicises ecology–economy–society interdependencies linking sustainability to social justice and eco-democracy (Rumpala, 2008). Political change is not left to the market place but must be governed through policy integration and participatory governing mechanisms. Environmental commitments must be transformed into constitutional norms in the same way as social welfare commitments, as part of a late-modern society: 'Environmental issues are seen as an outstanding manifestation of and challenge for a different modernity and a different political and social capacity for change and steering' (Arts *et al.*, 2006: 97). However, although this ideal-type narrative is transformationist, it is not consistently 'counter-hegemonic' (Swyngedouw, 2007). This is because not all the writers whose work can be grouped around this framing of interdependence consider economic development as necessarily against sustainability (Couturier and Thaimi, 2013). We return to this point in the fourth narrative below.

Sustainability as ecosystem approaches problematising dynamic ecology–society interdependencies

The third sustainability narrative presented here is sustainability as ecosystem approaches problematising dynamic ecology–society interdependencies. This narrative has its roots in ecology in the 1960s and 1970s, and in particular in new ways of understanding ecosystem responses to natural resource management and hence the impact of human activities on the environment (Holling, 1973; Odum, 1969). As with sustainability as politicised ecology–economy–society interdependencies, this narrative is not supported by a single theory, but rather by three main approaches of 'ecosystem-based management', 'ecosystem approach' and 'ecosystem services' (Waylen *et al.*, 2014).[11] Whereas these approaches hold in common a dynamic ecosystem focus to conceptualise sustainability, nonetheless their differences over ecology–economy–society hierarchies cause strong tensions at the heart of this narrative with potentially radically different governing solutions.

The theoretical origins of this third sustainability narrative lie in scholarly debates on the protection of the ecological integrity of ecosystems and biodiversity. Ecosystem means 'a dynamic complex of plant, animal and microorganism communities and their non-living environment interacting as a functional unit' (Secretariat of the Convention on Biological Diversity, 2004: 6; Odum, 1969; Tansley, 1935). Although up until the 1960s and 1970s conventional wisdom in ecology had considered ecosystems to be in a single 'steady state equilibrium', new experimental observations coupled with literature review and population modelling carried out by Holling (1973) enabled him to question this assumption. Instead, he found that the composition, structure and function of ecosystems appeared dynamic with multi-stable states (Folke, 2006). This was evidenced by

> complex non-linear relations between entities under continuous change and facing discontinuities and uncertainty from suites of synergistic stresses and shocks.
>
> (Folke *et al.*, 2002: 438)

Holling argued that research should be re-orientated around the 'resilience capacity' of ecosystems in multi-stable states, away from single steady state equilibrium frameworks:

> Resilience determines the persistence of relationships within a system and is a measure of the ability of these systems to absorb changes of state variables, driving variables, and parameters, and still persist.
>
> (Holling, 1973: 17)

Critically, these changing understandings within ecology did not limit ecosystems' frontiers to encompass only nature. This was to redress the erroneous 'assumption that human and natural systems can be treated independently' (Folke *et al.*, 2002: 438). Humans were considered to be integral parts of ecosystems:

> People cannot be separated from nature. Humans are fundamental influences on ecological patterns and processes and are in turn affected by them.
>
> (Grumbine, 1994: 31)

Defining ecosystems as 'interactions between biotic, abiotic and anthropic factors at different scales' (Carter *et al.*, 2016), scholars began to problematise what they now understood to be dynamic, complex and uncertain ecology–society interdependencies.

As with the previous two narratives on sustainability, sustainability as eco-system approaches problematising dynamic ecology–society interdependencies has carried with it a broad prescriptive agenda for governing these interdependencies, yet within which we can identify important differences between the types of instruments proposed (De Lucia, 2015; Waylen *et al.*, 2015). First, and within conservation biology, ecosystem-based theorising was mobilised by scholars to promote a radical shift in thinking about humans' role in nature, providing people

> the opportunity to reinterpret our place on the planet as one species among many. … Protecting ecological integrity becomes the ultimate test of whether people will learn to fit in with nature.
>
> (Grumbine, 1994: 34)

Adopting ecocentric approaches to the interdependence, these scholars hierarchised ecological issues. Scholars argued that land management's main goal was to protect the ecological integrity of ecosystems responding to the biodiversity crisis 'whilst *accommodating* human use' (Grumbine, 1994: 27; emphasis added). Arguments were made that effective management responses to changing understandings of ecosystems would be ones which would encourage restoration and less human use of resources in general. A strong role for public policy was advocated to protect biodiversity and create protected natural areas.[12] In these more ecocentric applications of ecosystem management, conservation biologists promoted the rewilding of nature and wilderness restoration projects in which protected natural areas would have limited and reduced human use (Grumbine, 1994; Noss, 1992).[13] These reforms were to be facilitated through citizen participation in environmental decision-making, which aimed at educating people on biodiversity loss, encouraging them to become advocates of environmental causes, and fostering environmental responsibility of resource use (Grumbine, 1994).

This biocentric approach within conservation biology, whilst introducing an important tension in this narrative, has not become the dominant approach espoused in a literature proposing change to tackle biodiversity loss. Instead, work in ecology problematising the ecology–society interface proved critical for the way in which this interdependence was ultimately taken up in international mobilisations on biodiversity through two other approaches: the ecosystem approach and ecosystem services. The 'ecosystem approach' was codified in the Convention on Biological Diversity (CBD), which was signed at the United Nations Conference on Environment and Development, 'the Rio Earth Summit' in 1992; the 'ecosystem services' concept was institutionalised as a scientific framework on a global scale through the Millennium Ecosystem Assessment (MEA) in 2005 (Norgaard, 2010).

Drawing on Holling's discoveries (1973), the work of Folke and colleagues in ecology (Berkes and Folke, 1998; Folke, 2006; Folke *et al.*, 2002) was particularly influential in this regard. Starting from the assumption that 'sustainability implies maintaining the capacity of ecological systems to support social and economic systems' (Berkes *et al.*, 2003: 2), Folke *et al.* argued that

> efforts should be made to create synergies between economic development, technological change and the dynamic capacity of the natural resource base to support social and economic development.
>
> (2002: 438)

Theorising now in terms of 'social-ecological systems' (Berkes and Folke, 1998), these scholars argued that this redefinition of human-nature relations necessitated a change in governance within a new prism of resilience. As they argued, resource management tools hitherto – for example, command and control and monocultural production approaches – had been premised upon positivist science and linear and mechanistic models of nature (Berkes *et al.*, 2003: 8–9).[14] This had resulted in an assumption that the societal impact on nature could be predicted. By contrast, the resilience school expected instead that ecosystem responses to human use would be non-linear, complex and uncertain (Folke *et al.*, 2002: 438). New understandings of ecosystem dynamics should thus lead to the development of new land and sea management tools; for example, on how to measure the environmental impacts of production practices (e.g. agriculture, fisheries, aquaculture), which should no longer be focused on single species effects nor measured in isolation either from one another or from their milieu (Folke *et al.*, 2002; Garcia, *et al.*, 2003; Odum, 1969). Such tools should be developed within new approaches to governance which would build flexibility into practices, allowing for knowledge gaps and scientific uncertainties on the environmental impacts of resource use. Calling this 'adaptive management', they intended this approach to facilitate

> learning by doing … wherein resource management policies can be treated as 'experiments' from within which managers can learn.
>
> (Berkes *et al.*, 2003: 9)

This would allow management to be flexible in the face of scientific uncertainty about ecosystem resilience (and later climate change; see Rocle and Salles, 2017).

These new conceptions of social-economic systems lie at the heart of the Ecosystem Approach whose international principles were set out in COP 5 Decision V/6 (the Malawi principles; Waylen *et al.*, 2014). In these international declarations, the Ecosystem Approach (EA) is defined as

> a strategy for the integrated management of land, water and living resources that promotes conservation and sustainable use in an equitable way.
>
> (www.cbd.int/ecosystem/)

Global definitions of the EA regard it as aimed at 'ecological well-being, human well-being and multi-sectoral integration' within a broader concern of social equity (Nunes *et al.*, 2011; Soto *et al.*, 2008). Although its conceptual expectations for changing governing institutions can vary between ecosystems compared (e.g. in forestry, Conchon *et al.*, 2015; in estuary, Bouleau and Boët, 2015; in marine, Carter *et al.*, 2015; Drouineau *et al.*, 2016), common expectations of changes to governance practices can be identified. These include: (i) an holistic approach to problem-solving aimed at grasping interconnections of the socio-ecosystem (often discussed in terms established by the DPSIR policy tool: Drivers, Pressures, Impact, State, Response; Bouleau and Pont, 2015; Lewison *et al.*, 2016); and (ii) adaptive and flexible governing arrangements across spatial and temporal scales (Gormley *et al.*, 2015).

Aiming at going beyond single species or single sector management, scholars detailing the EA consider it as endorsing integrative or transversal governance connecting different spaces of public action which would otherwise be disconnected from one another.[15] It is also an approach which has been associated with new forms of participatory governance (Leslie and McLeod, 2007; Van Leeuwen *et al.*, 2014; Waylen *et al.*, 2014). According to Waylen *et al.*, community-based approaches to resource management were influential in arguing for the value of local decision-making taken up in this approach. This was to ensure

> that communities and resource users can fully participate in environmental decision making that balances resource use with conservation.
>
> (2014: 1217)

Regulatory tools which it is claimed meet these objectives are: DPSIR; integrated ecosystem assessment; long-term management plans; integrated coastal zone management; and protected areas (Nunes *et al.*, 2011).

Running parallel to scholarly work on the EA, and cross-cutting it, conservation biologists, ecologists and environmental economists have elaborated the concept of ecosystem services (ES) (Costanza *et al.*, 1997; Daily, 1997).[16] Motivated by a realisation that the threat which ecosystem degradation posed not only to nature but also to humans' own sustainability was little appreciated in society (Daily, 1997), these scholars initially mobilised the concept of 'ecosystem services' to publicise 'the usefulness of nature for society, other than an object of ethical concern' (Braat and de Groot, 2012: 6):

> There was a strong sense that, however revolting for those who intrinsically value nature, the use of market metaphors was necessary to awaken a public deeply embedded in a global economy and distant from natural processes.
>
> (Norgaard, 2010)

Their audience was not only society in general, but especially conservation managers to enable them to make choices through making 'conservation economically attractive' (Daily and Matson, 2008). A first objective of this work was to name the services which ecosystems provided;[17] a second one was to set out methods for granting these services an economic value (for example, Costanza *et al.* [1997] estimated biodiversity loss to average US 33 trillion dollars per year). As argued by Salles, 'conservationists … hoped to find in economic analysis strong advocacy to stop biodiversity losses' (2011: 470), through the translation of 'the value of losses from the destruction of some ecosystems in terms that allow a comparison with other societal issues' (2011: 478). This was premised upon arguments that ecosystem services were being given too little weight in decision-making as neither were they quantified in economic terms nor were they 'captured' in commercial markets (Costanza *et al.*, 1997). This work was later described as follows:

> In order to make the economic value that nature provides visible, we need to estimate and disclose values for nature's goods and services (or so-called 'ecosystem services'). These estimated values can inform policy choices, executive actions, business decisions and consumer behavior.
>
> (TEEB, 2011: 3)

This enclosing of 'the natural world within the market world' (McAfee, 2012: 109) and commodification of nature as tradeable assets placed its emphasis on market-based instruments as effective governing tools to address biodiversity loss (TEEB, 2011). A number of such market-based instruments have been called for, including: biodiversity offsets, payments for ecosystem services, cost-benefit analysis, environmental accounting and performance systems (Paavola and Hubacek, 2013; TEEB, 2011). These could be used alongside other tools, such as environmental assessment, planning instruments and protected areas.

In summary, whereas this third narrative places analysis of the ecosystem (and social-ecosystem) at the heart of sustainability, it is nonetheless undercut by internal cleavages. Two key tensions can be observed with consequences for governing choices and the re-organisation of political power: the tension between ecocentric versus anthropocentric responses (De Lucia, 2015) to dynamic ecology–society interdependencies on the one hand, and the tension between distributional versus utilitarian visions of decision-making within anthropocentric approaches on the other. Mediations of the second tension therefore determine the consequences of this narrative for the re-organisation of state power, which will either be re-assigned or reduced. This will depend upon whether market-based instruments are being mooted as a more effective means for addressing biodiversity loss than state governing institutions (McAfee, 2012; Norgaard, 2010) or whether ES economic valuation takes place within broader projects also addressing distributional effects and common property rights issues (Lant *et al.*, 2008).

Sustainability as de-growth re-ordering ecology–society–economy hierarchies

The fourth narrative is sustainability as de-growth re-ordering ecology–society–economy hierarchies. This narrative has its roots in critical social and culturalist theory, ecological economics, ecosocialism and ecoMarxism (Boulding, 1966; Georgescu-Roegen, 1971; Gorz, 1978, 1991; Kapp, 1961; O'Conner, 1994; Pepper, 1998). More recently – since the early 2000s – it has been mobilised once more, this time as a social and political movement (Demaria *et al.*, 2013) and a critical approach within environmental sociology (Clausen and Clark, 2005; Longo, 2012). It is sustained by scientific work which is politicised in the sense that it is expressly 'counter-hegemonic' (Swyngedouw, 2014). This is because this narrative identifies key contradictions between the neoliberal industrial capitalist system and ecological protection (Asara, *et al.*, 2015; Pepper, 1998). This being said, and unlike extreme positions which argue that it is 'too late' for sustainability policies, this narrative does not reject outright the concept of sustainability (Asara *et al.*, 2015). However, scholars argue that what is urgently required is a re-hierarchising of ecology–society–economy interdependencies in favour of ecology and social justice. This is to be brought about through a downscaling of production and consumption. As argued, de-growth is not to be taken 'literally' or to be measured in GDP terms; rather, the aim is to effect a radical and redistributive political economic turn which is socially sustainable (Asara *et al.*, 2015; Gómez-Baggethun and Naredo, 2015).

At the heart of this narrative is an argument questioning whether industrial capitalist societies can ever be reformed to tackle environmental degradation and climate change or whether they are 'inherently unsustainable' (Lippert, 2010: 9):

> [N]eo-Marxists conceptualize the relationship of economic growth to environmental protection as a largely zero-sum game. No amount of technological innovation, green taxes, or ecological restructuring is perceived capable of rendering capitalism ecologically sustainable.
>
> (Davidson, 2012: 36)

Even if the state were to enact strong environmental policies, it is argued, the 'accumulation imperative' (Davidson, 2012: 36) of capitalist society will ultimately undermine any positive environmental effects witnessed because 'eco-efficiency gains are often re-invested in further consumption of economic activities that counterbalance the improvements achieved' (the so-called Jevons paradox; Asara *et al.*, 2015: 376). In this vein, scholars have argued that although potentially different meanings of sustainability exist, in practice and in international United Nations reports, dominant mantras in sustainability politics are growth *for* sustainability, market liberalisation and technocracy (Gomez-Baggethun and Naredo, 2015). Further, practices concretising these approaches have failed to address ecological degradation (Gomez-Baggethun and Naredo, 2015). Taking issue with the hypothesis that synergies can be found across ecological-economic-social interdependencies, the de-growth approach argues instead that 'harmonisation has proved elusive' (Asara *et al.*, 2015: 375) and that there is a fundamental contradiction between neoliberal growth and sustainable society (Swyngedouw, 2014). As such, a central objective of this narrative is to politicise both the 'post-political condition' of contemporary capitalism and its associated political status quo foreclosing alternatives (Swyngedouw, 2007).

Instead, what is required is

> far-reaching social change embodying fundamental principles of socialism, but melded with environmental goals based on stewardship.
>
> (Pepper, 1998: 2)

To counter dominant elements in the other sustainability narratives presented hitherto, this fourth narrative is consequently sustained by a political and scientific project which has become an anti-growth democratic movement since the early 2000s. It is a radical project and 'a confluence point where streams of critical ideas and political action converge' (Demaria *et al.*, 2013: 193).[18] The project advocates that mechanisms for achieving sustainability as de-growth can be multiple and allocates a transformationist role to both science and politics.

Change is expected to come about through 'intentional action' (Asara, *et al.*, 2015: 379), wherein the role of science is twofold. First, there is an expectation that through theorising and publishing work on sustainability as de-growth, scientists can offer an alternative interpretative frame of reference for political activists (Demaria *et al.*, 2013). This frame of reference can serve both as a slogan – i.e. naming actions already underway thereby giving them a collective political purpose (e.g. local movements advocating 'alternative transport') – and as a catalyst – i.e. sowing the seeds for different ways of living, thereby encouraging change towards dematerialisation. Second, science can research de-growth projects already in place, analyse them as case studies and publish their findings, thereby contributing further to the political project (e.g. on relocalised connected economies: Kunze and Becker, 2015). Consequently, a strong anticipated mechanism for change is oppositional activism, building alternatives through the setting up of 'local, decentralised, small-scale and participatory alternatives' (e.g. 'eco-villages, alternative (so-called ethical) banks or credit cooperatives, decentralised renewable energy cooperatives': Demaria *et al.*, 2013: 202). In this light, the importance of networks at the local and inter-communal scales is highlighted as critical for the governance of de-localised and dematerialised economies.

There are different points of view within this narrative in relation to the governance of reform more widely. Some authors consider that to bring about the kind of alternative society envisaged will require 'major shifts in national and supra-national political and economic structures' (Asara *et al.*, 2015: 381). Others, however, point to tensions between scholars inhabiting this narrative and an ambiguity in relation to the degree of institutional reform versus transformation deemed necessary:

> Another recurring debate is on the type of democratic system. On the one hand we might have to defend the democratic institutions put at risk with the economic crisis, and at the same time support the development of more participative ones. Similarly, while some take a traditional anarchist perspective in favour of abandoning the state, others believe the state should be kept and improved. In many cases, however, revolutionary positions could live together with reformist ones.
>
> (Demaria *et al.*, 2013: 203)

In summary, this narrative self-proclaims as political (as well as theoretical), making the case for a re-ordered hierarchy of ecology–society–economy interdependencies. Yet, as with the third narrative on sustainability as ecosystem approaches problematising dynamic ecology–society interdependencies, tensions within the fourth narrative mean that it is not possible to identify a clear direction in terms of state reform. Conflicts exist on whether or not sustainability as de-growth can be brought about within a transformed state.

Conclusions

Through revisiting a long-standing and a more recent sustainability literature, this chapter has discerned four 'ideal-type' sustainability narratives. Each narrative problematises sustainability in terms of a distinct set of interdependencies between nature and society and accordingly proposes change in the re-organisation of state power to govern interdependencies so defined. Working across the narratives we can observe cross-cutting similarities: for example, between depoliticised ecology–economy interdependencies and utilitarian versions of ecosystem services on the one hand, versus politicised ecology–economy–society interdependencies and territorially-embedded ecosystem approaches on the other. Or between those narratives aimed at politicising interdependencies (narratives 2 and 4) and those aimed at depoliticising them (narratives 1 and 3).[19]

As Table 2.1 makes clear, therefore, there are important differences both between and within these narratives. In particular, we can observe a central cleavage between those narrative elements which support an efficiency politics bringing about change to address problems of environmental degradation, and those which favour a distributive one. As is extensively argued within this literature, given that different actors are likely to support different narratives towards change, implementing sustainability will be a contested political project. Studying the narratives further confirms that this contest is

Table 2.1 Summary of 'ideal-type' sustainability narratives

Sustainability 'ideal-type' narrative	Lead disciplines	Re-organisation of state power	Policy tools	Theories of politics/key norms
De-politicised ecology–economy interdependencies	Environmental sociology; Political science	Reduced role for government acting on the market	Smart regulation; Market-based instruments; Environmental Impact Assessment	Utilitarian; Individualist; Efficiency; Economy–ecology win–win
Politicised ecology–economy–society interdependencies	Political science; Human geography; Environmental sociology	Strong role for environmental policy; Policy integration; Inclusive, participatory eco-democracy; Environmental protection as constitutional commitment	Environmental policy integration; Precautionary principle; Participatory eco-democracy; Green constitutional rights; Environmental Impact Assessment	Distributional; Communitarian (state scale); Equity; Economy–ecology–social justice
Ecosystem approaches problematising ecology–society interdependencies	Conservation biology; Ecology; Economics	Tensions between eco- versus anthropocentric ecosystem approaches; Tensions within anthropocentric approaches between change through re-assignment of state power and territory-building projects versus change through markets and reduced role for government	Wilderness and rewilding public programmes; Adaptive management; Integrated management, e.g. ICZM; Environmental Impact Assessment; DPSIR indicators; Participatory management; Market-based instruments	Tension: utilitarian/ distributional; Tension: individualist/ communitarian; Tension: efficiency/equity; Socio-ecosystem
De-growth re-ordering ecology–society–economy hierarchies	Critical cultural and social theory; Ecological economics; Environmental sociology	Opposition activism accompanied either by radical revolution (and abandonment) of the state or by state reform through participatory and/or direct democracy	Alternative economic instruments (e.g. credit cooperatives); Participatory/direct democracy	Distributional; Communitarian (local scale); Equity; Ecology–social justice

not only over whether or not to address ecological degradation (and climate change) – although it is still that – but is clearly related to the 'values' (Smith, 2016) underpinning sustainability and the consequent form of political organisation and governing arrangements advocated (Wright and Kurian, 2009).

These observations must be taken in relation to a secondary point, namely that it is important not to reify these narratives (Bevir and Rhodes, 2003). For the purposes of comprehension and method, I have identified these narratives based on a wide literature review. However, of course, they only exist in practice once enacted by actors. We therefore need empirical research to find out which of these sustainability narratives is giving meaning to the governing of European fish farming aquaculture (see Chapters 4, 5 and 6).

Presenting this literature in terms of narratives not only serves to extend our understanding of sustainability, it also helps us methodologically. Central elements inherent in each narrative can be used to code research findings and enable us to categorise the institutional and political work of actors, potentially revealing shifts in the uptake of different narratives over different periods of time (I return to this point in Chapter 3). The approach via narratives is particularly useful given that categorising 'the governing of sustainability' solely on the basis of the existence of a particular kind of policy instrument can be misleading. This point is well illustrated in the case of the policy tool of Environmental Impact Assessment: the first three narratives all endorse Environmental Impact Assessments but conceptualise these policy tools in different ways concerning 'what' is expected to be measured in the course of an assessment, at 'which scale', and 'how' this policy instrument is expected to work in conjunction with others.

Finally, each sustainability narrative promotes political change. However, sustainability is not the only imperative towards political change. Other imperatives exist, which will also come to shape how actors position themselves, both in respect to different sustainability narratives and also the degree to which they are successful in institutionalising one of these narratives at the expense of the others. It is to this specific issue which we will now turn in Chapter 3.

Notes

1 This shift in the focus of inquiry from 'sustainability' to 'interdependencies and their narratives' holds methodological consequences which I discuss in the last section of Chapter 3.
2 Thinking in terms of new interdependencies also underpinned the much-cited work of Meadows *et al.* (1972) and their report on *The Limits to Growth*. In this study, a dynamic computer model was run both to simulate different interactions between what they termed the five sub-systems of the global system (population growth, food production, industrial production, pollution, and consumption of non-renewable natural resources) and also to set out scenarios for global sustainability dependent on different combinations of these aforementioned interdependencies.

3 According to conventional wisdom, 'Brundtland' is the defining moment in sustainability's history as a global governing concept, even if scholars have identified international formulations of sustainability which pre-date this report. For example, Langhelle (2000) points to the 1972 report by a working group within the World Council of Churches which speaks of a sustainable society. Hopwood *et al.* state that 'the first important use of the term was in 1980 in the World Conservation Strategy (IUCN *et al.*, 1980)' (2005: 39).

4 The Brundtland Report has been criticised for these very reasons; see Asara *et al.*, 2015; Dobson. 1998; Gómez-Baggethun and Naredo, 2015. As argued by Springett (2013), the Report placed its emphasis on poverty, not limits to growth. It also concentrated on the global North–South divide. Growth was kept on the table because it was seen as important for developing countries.

5 Ecological modernisation was founded by Joseph Huber, an environmental sociologist, and Martin Jänicke, a political scientist.

6 This optimistic and deterministic view of technology (York and Rosa, 2003) was later further explained as involving two orders of technological innovation. The first was to reduce the environmental impact of both production practices and products in existing industries, focusing on 'greening' and 'efficiency' (Huber, 2000; Hajer, 2005 [1995]). The second was what Hajer called 'structural change', and Huber a 'strategy of consistency', i.e. new kinds of technological innovation 'to change the qualities of the industrial metabolism' (Huber, 2000: 281) and potentially replace old industrial practices with new industries – e.g. organic farming, solar energy, GM biotechnology (Huber, 2000). In his early work, Jänicke emphasised that technological innovation was not just about resource efficiency, but required structural change of key sectors such as construction, road and transport.

7 The frontiers between ecological modernisation theory, sustainable development, and environmental governance literature are disputed. For some scholars, 'ecological modernisation is even a synonym of sustainable development' (Lippert, 2010: 8). For these, 'ecological modernisation theory suggests that 'modernity' enters the stage of sustainable development through ecological modernisation' (Lippert, 2010: 11). For others, ecological modernisation, in both its versions, is quite distinct from sustainable development (see Langhelle, 2000).

8 For example, Haughton (1999) set out five framings of equity within sustainability: futurity – inter-generational equity; social justice – intra-generational equity; transfrontier responsibility – geographical equity; people treated openly and fairly – procedural equity; and importance of biodiversity – inter-species equity.

9 Rio Declaration and Agenda 21 are international agreements consolidating participatory sustainable development policy and indicate the uptake of this narrative on a global scale.

10 For Eckersley, ecologically informed counter developments are environmental multilateralism, ecological modernisation and the emergence of green discursive designs. Her evocation of ecological modernisation as a counter development has left her open to ecosocialist critique (see Davidson, 2012).

11 As we saw with the fault lines between sustainable development and ecological modernisation, there are terminology and boundary issues within and between ecosystem approaches. Some scholars use the term 'ecosystem approach' as a strategy which encompasses all three approaches named in this chapter, and indeed this is how the ecosystem approach is defined by the Convention on Biological Diversity Guidelines (Secretariat of the Convention on Biological Diversity, 2004). Yet others argue that the 'ecosystem approach' differs from both 'ecosystem-based management' and 'ecosystem services' in important respects (Waylen *et al.*, 2014). I have therefore used the term 'ecosystem approaches' in

the plural to encompass all three approaches, one of which I refer to as the 'eco-system approach'.

12 In some cases, arguments were made for strong public policy to reduce human population growth (see Grumbine, 1994: 33).

13 For example, Noss argued that the re-introduction of Panthers in the National Forests of Florida Project would require the closing of roads, the curtailing of logging and the ceasing of recreational activities: 'The best way to accomplish most of the necessary changes would be to designate most of each Forest as wilderness' … 'so that Panthers can follow their normal instincts to wander without coming into frequent contact with humans' … 'none of the terrestrial ecosystems of Florida can be considered complete until they regain healthy populations of their top predator, the Panther' (Noss, 1992: 21–22).

14 See Carter (2013) for an account of these arguments in the changing government of EU fisheries.

15 For example, the application of the Ecosystem Approach in the EU Marine Strategy Framework Directive seeks to address all possible human impacts on marine waters, including, *inter alia*, fisheries, dredging, contaminants, noise, litter, etc. (Bouleau *et al.*, 2018; Van Leeuwen *et al.*, 2014).

16 The epistemological history of ecosystem services can be traced to these central works: Walter Westman, 1977; Farnworth *et al.*, 1981; Costanza *et al.*, 1997; Daily, 1997.

17 Ecosystem services (ES) are defined as nature's goods and services that benefit society: e.g. storage of carbon by soils, vegetation and oceans, habitats for plants, animals and micro-organisms, filtering of fresh water, food for production (e.g. fish) and even the aesthetic or spiritual significance of landscapes (MEA, 2005). The MEA (Millennium Ecosystem Assessment) categorised these into four main groups of services: supporting, regulating, provisioning and cultural services.

18 For example, one of the founding members of the de-growth movement, François Schneider, toured France on a donkey for over a year 'spreading the proposal of degrowth' (Demaria, 2013: 193).

19 These cross-cutting synergies and differences also reflect tensions within disciplines on ontological and epistemological questions.

References

Arts, B., Leroy, P. Tatenhove, J. 2006. 'Political modernisation and policy arrangements: a framework for understanding environmental policy change', *Public Organization Review*, 6: 93–106.

Asara, V., Otero, I., Demaria, F., Corbera, E. 2015. 'Socially sustainable degrowth as a social–ecological transformation: repoliticizing sustainability', *Sustainability Science*, 10: 375–384.

Baker, S., Kousis, M., Richardson, D., Young, S. 1997. *The Politics of Sustainable Development: Theory, Policy and Practice within the European Union*, London: Routledge.

Barbier, R., Larrue, C. 2011. 'Démocratie environnementale et territoires: un bilan d'étape', *Participations*, 1(1): 67–104.

Becker, E., Jahn, T., Stiess, I. 1999. 'Exploring uncommon ground: sustainability and the social sciences'. In Becker, E. and Jahn, T. (eds) *Sustainability and the Social Sciences: A Crossdisciplinary Approach to Integrating Environmental Considerations into Theoretical Reorientation*, London: Zed, pp. 1–22.

Berkes, F., Folke, C. (eds). 1998. *Linking Social and Ecological Systems: Management Practices and Social Mechanisms for Building Resilience*, Cambridge: Cambridge University Press.

Berkes, F., Colding, J., Folke, C. 2003. *Navigating Social–Ecological Systems: Building Resilience for Complexity and Change*, Cambridge: Cambridge University Press.

Bevir, M., Rhodes, R. 2003. *Interpreting British Governance*, Routledge, London.

Blackstock, K. Richards, C. 2007. 'Evaluating stakeholder involvement in river basin planning: a Scottish case study', *Water Policy*, 9: 493–512.

Boulding, K. 1966. 'The economics of the coming spaceship Earth'. In Jarrett, H. (ed.) *Environmental Quality in a Growing Economy*, Baltimore, MD: Resources for the Future/Johns Hopkins University Press, pp. 3–14.

Bouleau, G., Boët, P. 2015. 'L'approche écosystémique pour les estuaires et les cours d'eau'. Paper presented at ECOGOV project meeting, 4 June, INRA Pierroton.

Bouleau, G., Pont, D. 2015. 'Did you say reference conditions? Ecological and socio-economic perspectives on the European Water Framework Directive', *Environmental Science & Policy*, 47: 32–41.

Bouleau, G., Carter, C., Thomas, A. 2018 forthcoming. 'Suivre les médiations entre connaissances et décisions dans les dispositifs participatifs de gestion de l'eau: comparaison de l'application de la DCE et de la DCSMM', *Participations*.

Braat, L., de Groot, R. 2012. 'The ecosystem services agenda: bridging the worlds of natural science and economics, conservation and development, and public and private policy', *Ecosystem Services*, 1(1): 4–15.

Brooks, D. 1992. 'The challenge of sustainability: is integrating environment and economy enough?', *Policy Sciences*, 25: 401–408.

Buttel, F. 2000. 'Ecological modernization as social theory', *Geoforum*, 31: 57–65.

Carson, R. 1962. *Silent Spring*, Penguin Books: London.

Carter, C. 2013. 'Constructing sustainability in EU fisheries: re-drawing the boundary between science and politics?', *Environmental Science and Policy*, 30: 26–35.

Carter, C., Cazals, C.. Hildermeier, J., Michel, L., Villareal, A. 2014. 'Sustainable development policy: "Competitiveness" in all but name'. In Jullien, B. and Smith, A. (eds) *The EU's Government of Industries: Markets, Institutions and Politics*, Abingdon: Routledge, pp. 165–189.

Carter, C., Salles, D., Caill-Milly, N., Morandeau, G., Auby, I., Oger Jeanneret, H. 2015. 'L'approche écosystémique et la gestion des écosystèmes marins'. Paper presented at ECOGOV project meeting, 4 June, INRA Pierroton.

Carter, C., Thomas, A., Salles, D., Caill-Milly, N., Morandeau, G., Auby, I., Oger Jeanneret, H. 2016. 'New integrated framework to grasp transforming science-politics "coupling practices": participatory European coastal and marine water management'. Poster presented at Estuarine Coastal Sciences Association conference, 4–7 September, Bremen.

Christoff, P. 1996. 'Ecological modernisation, ecological modernities', *Environmental Politics*, 5(3): 476–500.

Clausen, R., Clark, B. 2005. 'The metabolic rift and marine ecology: an analysis of the ocean crisis within capitalist production', *Organization and Environment*, 18(4): 422–444.

Conchon, P., Carnus, J-M., Sergent, A. 2015 'L'approche écosystémique et la gestion des écosystèmes forestiers'. Paper presented at ECOGOV project meeting, 4 June, INRA Pierroton.

Costanza, R., d'Arge, R., de Groot, R., Farber, S., Grasso, M., Hannon, B., Limburg, K., Naeem, S., O'Neill, R., Paruelo, J., Raskin, R., Sutton, P., van den Belt, M. 1997. 'The value of the world's ecosystem services and natural capital', *Nature*, 387: 253–260.

Couturier, A., Thaimai, K, 2013. 'Eating the fruit of the poisonous tree? Ecological modernisation and sustainable consumption in the EU'. Institute for International Political Economy Working Paper 20/2013, Berlin, pp. 1–27.

Daily, G. (ed.) 1997. *Nature's Services: Societal Dependence on Natural Ecosystems*, Washington: Island Press.

Daily, G., Matson, P. 2008. 'Ecosystem services: from theory to implementation', *PNAS*, 105(28): 9455–9456.

Daly, H. 1973. *Toward a Steady-State Economy*, San Francisco: W.H. Freeman.

Davidson, S. 2012. 'The insuperable imperative: a critique of the ecologically modernizing state', *Capitalism Nature Socialism*, 23(2): 31–50.

De Lucia, V. 2015. 'Competing narratives and complex genealogies: the ecosystem approach in international environmental law', *Journal of Environmental Law*, 27: 91–117.

Deldrève, V., Candau, J. 2014. 'Produire des inégalités environnementales justes?', *Sociologie*, 3(5): 255–269.

Demaria, F., Schneider, F., Sekulova, F., Martinez-Alier, J. 2013. 'What is degrowth? From an activist slogan to a social movement', *Environmental Values*, 22: 191–215.

Dobson, A. 1996. 'Environmental sustainabilities: an analysis and a typology', *Environmental Politics*, 5: 401–428.

Dobson, A. 1998. *Justice and the Environment*, Oxford: Oxford University Press.

Drouineau, H., Lobry, J., Bez, N., Travers-Trolet, M., Vermard, Y., Gascuel1, D. 2016. 'The need for a protean fisheries science to address the degradation of exploited aquatic ecosystems', *Aquatic Living Resources*, 29(E201): 1–7.

Dryzek, J. 1997. *The Politics of the Earth: Environmental Discourses*, New York: Oxford University Press.

Duit, A., Feindt, P., Meadowcroft, J. 2016. 'Greening Leviathan: the rise of the environmental state?', *Environmental Politics*, 25(1): 1–23.

Eckersley, R. (ed.). 1995. *Markets, the State, and the Environment: Towards Integration*. South Melbourne, Australia: Macmillan Education.

Eckersley, R. 2004. *The Green State: Rethinking Democracy and Sovereignty*, Cambridge MA: MIT Press.

European Environment Agency. 2001. *Late Lessons from Early Warnings: The Precautionary Principle 1896–2000* Environmental Issue Report No. 22 Luxembourg: Office for Official Publications of the European Communities.

Farnworth, E., Tidrick, T., Jordan, C., Smathers, W. 1981. 'The value of natural ecosystems: an economic and ecological framework', *Environmental Conservation*, 8(4): 275–282.

Faucher, F. 1999. 'Party organisation and democracy: a comparison of les Verts and the British Green Party', *GeoJournal* 47: 487–496.

Folke, C. 2006. 'Resilience: the emergence of a perspective for social–ecological systems analyses', *Global Environmental Change*, 16(3): 253–267.

Folke, C., Carpenter, S., Elmqvist, T., Gunderson, L., Holling, C., Walker, B. 2002. 'Resilience and sustainable development: building adaptive capacity in a world of transformations', *AMBIO: A Journal of the Human Environment*, 31(5): 437–440.

Garcia, S.M., Zerbi, A., Aliaume, C., Do Chi, T., Lasserre, G. 2003. 'The ecosystem approach to fisheries. Issues, terminology, principles, institutional foundations, implementation and outlook'. FAO Fisheries Technical Paper no. 443, Rome: Food and Agriculture Organisation (FAO).

Georgescu-Roegen, N. 1971. *The Entropy Law and the Economic Process*, Cambridge MA: Harvard University Press.

Giddings, B., Hopwood, B., O'Brien, G. 2002. 'Environment, economy and society: fitting them together into sustainable development', *Sustainable Development*, 10: 187–196.

Gómez-Baggethun, E., Naredo, J.M. 2015. 'In search of lost time: the rise and fall of limits to growth in international sustainability policy', *Sustainability Science*, 10(3): 385–395.

Gormley, K., Hull, A., Porterac, J., Bell, M., Sanderson, W. 2015. 'Adaptive management, international co-operation and planning for marine conservation hotspots in a changing climate', *Marine Policy*, 53: 54–66.

Gorz, A. 1978. *Écologie et Politique*, Paris: Editions du Seuil.

Gorz, A. 1991. *Capitalisme, Socialisme, Écologie*, Paris: Editions Galilée.

Grabosky, P.N. 1995. 'Governing at a distance: self-regulating green markets'. In Eckersley, R. (ed.) *Markets, the State, and the Environment: Towards Integration*. South Melbourne, Australia: Macmillan Education, pp. 197–228.

Greenwood, D. 2015. 'In search of Green political economy: steering markets, innovation, and the zero carbon homes agenda in England', *Environmental Politics*, 24(3): 423–441.

Grumbine, R.E. 1994. 'What is ecosystem management?', *Conservation Biology*, 8(1): 27–38.

Hajer, M. 2005 [1995]. *The Politics of Environmental Discourse: Ecological Modernization and the Policy Process*, 2nd edition, Oxford: Oxford University Press.

Haughton, G. 1999. 'Environmental justice and the sustainable city', *Journal of Planning Education and Research*, 18: 233–243.

Haughton, G., Counsell, D. 2004. *Regions, Spatial Strategies and Sustainable Development*, London: Routledge.

Hay, C. 2010. 'Introduction: political science in an age of acknowledged interdependence'. In Hay, C. (ed.) *New Directions in Political Science: Responding to the Challenges of an Interdependent World*, Basingstoke: Palgrave, pp. 1–24.

Holling, C.S. 1973. 'Resilience and stability of ecological systems', *Annual Review of Ecology and Systematics*, 4: 1–23.

Holmes, T., Scoones, I. 2000. 'Participatory environmental policy processes: experiences from North and South'. Working Paper 113, Institute of Development Studies, University of Sussex, Brighton, Sussex.

Hopwood, B., Mellor, M., O'Brien, G. 2005. 'Sustainable development: mapping different approaches', *Sustainable Development*, 13: 38–52.

Horlings, L., Marsden, T. 2011. 'Towards the real green revolution? Exploring the conceptual dimensions of a new ecological modernisation of agriculture that could "feed the world"', *Global Environmental Change*, 21(2): 441–452.

Horlings, L., Marsden, T. 2014. 'Exploring the "new rural paradigm" in Europe: eco-economic strategies as a counterforce to the global competitiveness agenda', *European Urban and Rural Studies*, 21(1): 4–20.

Huber, J. 2000. 'Towards industrial ecology: sustainable development as a concept of ecological modernization', *Journal of Environmental Policy & Planning*, 2(4): 269–285.

IUCN, UNEP, WWF. 1980. *World Conservation Strategy: Living Resource Conservation for Sustainable Development*, Gland, Switzerland: IUCN.

Jänicke, M., Lindemann, S. 2010. 'Governing environmental innovations', *Environmental Politics*, 19(1): 127–141.

Jordan, A. 2008. 'The governance of sustainable development: taking stock and looking forwards', *Environment and Planning C: Government and Policy* 26: 17–33.

Jordan, A., Adelle, C. (eds). 2013. *Environmental Policy in the EU: Actors, Institutions and Processes*, 3rd edition, Abingdon: Routledge.

Jordan, A., Lenschow, A. (eds). 2008. *Innovation in Environmental Policy? Integrating the Environment for Sustainability*. Cheltenham: Edward Elgar.

Jordan, A., Lenschow, A. 2010. 'Environmental policy integration: a state of the art review', *Environmental Policy and Governance*, 20(3): 147–158.

Jouzel, J.N., Dedieu, F. 2013. 'Rendre visible et laisser dans l'ombre. Savoir et ignorance dans les politiques de santé au travail', *Revue française de science politique*, 1(63): 29–49.

Kapp, K.W. 1961. *Toward a Science of Man in Society: A Positive Approach to the Integration of Social Knowledge*, The Hague: Martinus Nijhoff.

Kunze, C., Becker, S. 2015. 'Collective ownership in renewable energy and opportunities for sustainable degrowth', *Sustainability Science* 10(3): 425–437.

Lafferty, W., Hovden, E. 2003. 'Environmental policy integration: towards an analytical framework', *Environmental Politics*, 12(3): 1–22.

Lange, P., Driessen, P., Sauer, A., Bornemann, B., Burger, P. 2013. 'Governing towards sustainability-conceptualizing modes of governance', *Journal of Environmental Policy & Planning*, 15(3): 403–425.

Langhelle, O. 2000. 'Why ecological modernization and sustainable development should not be conflated', *Journal of Environmental Policy & Planning*, 2(4): 303–322.

Lant, C., Ruhl, J., Kraft, S. 2008. 'The tragedy of ecosystem services', *BioScience*, 58(10): 969–974.

Lenschow, A. (ed.) 2002. *Environmental Policy Integration: Greening Sectoral Policies in Europe*, London: Earthscan.

Leroy, P., Van Tatenhove, J. 2000. 'Political modernization theory and environmental politics'. In Spaargaren, G., Mol, A. and Buttel, F. (eds) *Environment and Global Modernity*, London: Sage, pp. 187–208.

Leslie, H., McLeod, K. 2007. 'Confronting the challenges of implementing marine ecosystem-based management', *Frontiers in Ecology and the Environment*, 5(10): 540–548.

Lewison, R., Rudd, M., Al-Hayek, W., Baldwin, C., Beger, M., Lieske, S., Jones, C., Satumanatpan, S., Junchompoo, C., Hines, E. 2016. 'How the DPSIR framework can be used for structuring problems and facilitating empirical research in coastal systems', *Environmental Science and Policy*, 56: 110–119.

Lippert, I. 2010. *Agents of Ecological Modernisation*. Tönning, Lübeck, Marburg: Der Andere Verlag.

Longo, S. 2012. 'Mediterranean rift: socio-ecological transformations in the Sicilian bluefin tuna fishery', *Critical Sociology*, 38(3): 417–436.

MacGregor, S. 2004. 'From care to citizenship: calling ecofeminism back to politics', *Ethics and the Environment*, 9(1): 56–84.

McAfee, K. 2012. 'The contradictory logic of global ecosystem services markets', *Development and Change*, 43(1): 105–131.

MEA (Millennium Ecosystem Assessment). 2005. *Ecosystems and Human Well-being: Synthesis*. Washington (DC): Island Press.

Meadows, D.H., Meadows, D.L., Randers, J., Behrens, W. 1972. *The Limits to Growth*, London: Universe Books.

Norgaard, R. 2010. 'Ecosystem services: from eye-opening metaphor to complexity blinder', *Ecological Economics*, 69: 1219–1227.

Noss, I.Z. 1992. 'The wildlands project: land conservation strategy', *Wild Earth*, 1: 10–25.

Nunes, J.P., Ferreira, J.G., Bricker, S.B., O'Loan, B., Dabrowski, T., Dallaghan, B., Hawkins, A.J.S., O'Connor, B., O'Carroll, T. 2011. 'Towards an ecosystem approach to aquaculture: assessment of sustainable shellfish cultivation at different scales of space, time and complexity', *Aquaculture*, 315: 369–383.

O'Connor, J. (ed.) 1994. *Is Capitalism Sustainable? Political Economy and the Politics of Ecology*, London: Guilford Press.

Odum, E.P. 1969. 'The strategy of ecosystem development', *Science*, 164: 262–270.

Paavola, J., Hubacek, K. 2013. 'Ecosystem services, governance, and stakeholder participation: an introduction', *Ecology and Society*, 18(4): 42.

Pepper, D. 1998. 'Sustainable development and ecological modernization: a radical homocentric perspective', *Sustainable Development*, 6: 1–7.

Pestre, D. 2011. 'Développement durable: anatomie d'une notion', *Natures Sciences Sociétés* 1(19): 31–39.

Reed, M. 2008. 'Stakeholder participation for environmental management: a literature review', *Biological Conservation*, 141: 2417–2431.

Rocle, N., Salles, D. 2017. 'Pioneers but not guinea pigs: experimenting with climate change adaptation in French coastal areas', *Policy Sciences*: 1–17.

Rudd, M. 2014 'Scientists' perspectives on global ocean research priorities', *Frontiers in Marine Science*. Available at: http://journal.frontiersin.org/Journal/10.3389/fmars.2014.00036/abstract.

Rumpala, Y. 2008. 'Le "développement durable" appelle-t-il davantage de démocratie? Quand le "développement durable" rencontre la "gouvernance"', *VertigO – la revue électronique en sciences de l'environnement* [online], 8(2): 1–20.

Salles, J.-M. 2011. 'Valuing biodiversity and ecosystem services: why put economic values on nature?', *Comptes Rendus Biologies*, 334: 469–482.

Scoones, I. 2016. 'The politics of sustainability and development', *Annual Review of Environment and Resources*, 41: 293–319.

Secretariat of the Convention on Biological Diversity. 2004. *The Ecosystem Approach* (CBD Guidelines), Montreal.

Smith, A. 2016. *The Politics of Economic Activity*, Oxford: Oxford University Press.

Soto, D., Aguilar-Manjarrez, J., Brugère, C., Angel, D., Bailey, C., Black, K., Edwards, P., Costa-Pierce, B., Chopin, T., Deudero, S., Freeman, S., Hambrey, J., Hishamunda, N., Knowler, D., Silvert, W., Marba, N., Mathe, S., Norambuena, R., Simard, F., Tett, P., Troell, M., Wainberg, A. 2008. 'Applying an ecosystem-based approach to aquaculture: principles, scales, and some management measures'. In Soto, D., Aguilar-Manjarrez, J. and Hishamunda, N. (eds) *Building an Ecosystem Approach to Aquaculture*, FAO/Universitat de les Illes Balears Experts Workshop. 7–11 May 2007, Palma de Mallorca, Spain. FAO Fisheries and Aquaculture Proceedings No. 14, Rome: Food and Agriculture Organisation (FAO), pp. 15–35.

Spaargaren, G. 2000. 'Ecological modernization theory and the changing discourse on environment and modernity'. In Spaargaren, G., Mol, A. and Buttel, F. (eds) *Environment and Global Modernity*, London: Sage, pp. 41–71.

Springett, D. 2013. 'Editorial: critical perspectives on sustainable development', *Sustainable Development*, 21: 73–82.

Swyngedouw, E. 2007. 'Impossible/undesirable sustainability and the post-political condition'. In Krueger, J.R. and Gibbs, D. (eds) *The Sustainable Development Paradox*, New York: Guilford Press, pp. 13–40.

Swyngedouw, E. 2014. 'Depoliticization ("the political")'. In D'Alisa, G., Demaria, F. and Kallis, G. (eds) *Degrowth: A Vocabulary for a New Era*. London: Routledge, pp. 90–93.

Tansley, A.G. 1935. 'The use and abuse of vegetational concepts and terms', *Ecology*, 16: 284–307.

TEEB (The Economics of Ecosystems and Biodiversity). 2011. *TEEB for Local and Regional Policy Makers*, London: Earthscan.

Toke, D. 2011. *Ecological Modernization and Renewable Energy*, Basingstoke: Palgrave.

Van Leeuwen, J., Raakjaer, J., Van Hoof, L., Van Tatenhove, J., Long, R., Ounanian, K. 2014. 'Implementing the marine strategy framework directive: a policy perspective on regulatory, institutional and stakeholder impediments to effective implementation', *Marine Policy*, 50(B): 325–330.

Van Tatenhove, J., Leroy, P. 2003. 'Environment and participation in a context of political modernisation', *Environmental Values*, 12(2): 155–174.

Waylen, K., Hastings, E., Banks, E., Holstead, K., Irvine, R., Blackstock, K. 2014. 'The need to disentangle key concepts from ecosystem-approach jargon', *Conservation Biology*, 28(5): 1215–1224.

Waylen, K., Blackstock, K., Holstead, K. 2015. 'How does legacy create sticking points for environmental management? Insights from challenges to implementation of the ecosystem approach', *Ecology and Society*, 20(2): 21.

WCED (World Commission on Environment and Development). 1987. *Our Common Future* (Brundtland Report), Oxford: Oxford University Press.

Weale, A., Williams, A. 1992. 'Between economy and ecology? The single market and the integration of environmental policy', *Environmental Politics*, 1(4): 45–64.

Westman, W. 1977. 'How much are nature's services worth?', *Science*, 197(4307): 960–964.

Williamson, D., Lynch-Wood, G. 2012. 'Ecological modernisation and the regulation of firms', *Environmental Politics*, 21(6): 941–959.

Wright, J., Kurian, P. 2009. 'Ecological modernization versus sustainable development: the case of genetic modification regulation in New Zealand', *Sustainable Development*, 18(6): 398–412.

York, R., Rosa, E. 2003. 'Key challenges to ecological modernization theory', *Organization and Environment*, 16(3): 273–288.

3 From sustainability to sustainability interdependence

Recomposing sustainability in a new analytical framework

Introduction

As I have both argued and demonstrated in Chapter 2, there are different genealogies and narratives of sustainability within a rich scientific literature spanning several decades and disciplines. Clearly, these different sustainability narratives of nature–society interdependencies provide actors governing European fish farming aquaculture with a rich range of possible theories, identities, instruments and stories (Bevir, 2011; Bevir and Rhodes, 2003) which can be mobilised by them to govern the issues facing this industry as regards sustainability. However, before I go on to examine how these different narratives are being enacted (or not) in European fish farming aquaculture, I need first to set out how I intend to grasp the politics of their implementation. In this vein, a central ambition in this chapter is to recompose 'sustainability' analytically through integrating it to a wider theory of political change and a dynamic institutionalist framework of sustainability interdependence.

The chapter is organised in three sections. In the first, I explain the need for a new framework. Clearly, sustainability and interdependence are concepts which have long since been associated with one another. But is the nature–society power dynamic the only way of understanding interdependence in the context of sustainability? Might there not be additional forms of interdependence at work which impact on actors' governing choices with broader political causes and consequences? Drawing upon long-standing premises consolidated within the public policy analysis literature highlighting the importance of policy–polity dialectics in public action transformation as well as in the governance of industry (Carter and Smith, 2008, 2009; Douillet, 2003; Itçaina et al., 2016; Jullien and Smith, 2014; Lagroye, 1991; Muller, 2015; Ollitrault, 2004; Revel et al., 2007 Richardson, 1996; Van Tatenhove, 2013; Wallace, 2000, Webb, 1977), I contend that sustainability-induced governing change does not take place in an 'institutional void' (Hajer, 2003). Rather, it takes place within broader processes of *already on-going* political change, whose causal factors are subject to rich debates.

Consequently, in a second section, I proceed to widen the parameters of my inquiry. More specifically, I set out the contents and politics of three on-going governing transformations distributing and re-distributing power across the political system and giving rise to new interdependence characteristics. These are: (i) interdependence and territory; (ii) interdependence and regulation; and (iii) interdependence and knowledge. These interdependencies have emerged following *inter alia* empirical and political projects on legitimate European economic governance, political debates on territorial identity, state regulatory reform and democratisation of knowledge use in decision-making. At times, their causes also stem from action to give effect to sustainability and I highlight this in each case.

Following from these discussions, in a third and final section, I expressly reframe the research object as sustainability interdependence and explain my conceptual and analytical tools. This is done through defining sustainability as an institution and public action principle, and integrating it within a wider theory of the political sociology of institutions.

In proceeding thus, it is important to restate my ontological positioning concerning sustainability and governance more generally. This matters because, as Chapter 2 attests, the sustainability literature on governance is especially charged with published work documenting what Farrell *et al.* have described as 'governance for' sustainability (2005). This prescriptive positioning has often been facilitated by its being underpinned (often implicitly) by theories of change operating from within a rationalist 'exclusive ontology of institutions' (Kauppi, 2010):[1]

> Despite this diversity the public administration science perspective predominates, with the suggestion of a unilinear trend from rational, hierarchical steering and pursuit of policy to governance, network steering and contextual pursuit of policy.
>
> (Arts *et al.*, 2006: 95)

This has given rise to formal sustainability institutionalist frameworks to assess whether particular governing arrangements will automatically promote particular forms of sustainability (Christen and Schmidt, 2012; Lange *et al.*, 2013). This is not the kind of institutionalist framework which is being developed here. Rather, my overarching aim is to develop a framework which permits a critical analysis of *actual* governing practices. The mobilisation of a political sociology of institutions to theorise this framework is coherent with this objective and finds its ontological home with other work on sustainability using sociological and constructivist institutionalist theories. Indeed, early on the prescriptive trend in sustainability analysis had already led scholars to argue that

> theories which make some separation between the process of achieving sustainability versus the content or actual details of what sustainability is deemed to be by society will avoid some of the accusations that we are building prescriptive models.
>
> (Jennings and Zandbergen 1995: 1023)

For Jennings and Zandbergen, these theories were ones of sociological institutionalism. This was to mark the beginning of an important strand of work within the alternative 'governance and' sustainability approach (Farrell *et al.*, 2005)[2] studying the implementation of sustainability in different places and contexts. In these collective publications, sociological and constructivist institutional approaches have examined 'sustainability and governance', explaining diversity in causes and trajectories of sustainability public action (Behagel and Turnhout, 2011; Boezeman, 2015; Bouleau *et al.*, 2009; Candau *et al.*, 2012; Cleaver and Franks, 2005; Fortin *et al.*, 2010; Jennings and Zandbergen, 1995; Sergent, 2014; Wesselink *et al.*, 2013). Through mobilising a political sociology of institutions to theorise 'sustainability interdependence', my case study will ultimately contribute to these debates, producing new knowledge on European fish farming aquaculture.

The need for a new framework

A long-standing tenet of public policy analysis is that the relationship between the governing of public problems and the reproduction of the polity is a critical one to grasp in explanations of policy (and political) change (Carter and Smith, 2008, 2009; Douillet, 2003; Itçaina *et al.*, 2016; Jullien and Smith, 2014; Lagroye, 1991, Muller, 2015; Ollitrault, 2004; Revel *et al.*, 2007; Richardson, 1996; Van Tatenhove, 2013; Wallace, 2000, Webb, 1977). Through analysing policy processes, research has not only revealed findings about public policy per se, but has also provided knowledge about polity-creating processes in which policy choices are made. This is to understand that actors and institutions interact across different territorial, sectoral and temporal scales of governance (Salles, 2009), whereby public action to implement particular policies will at the same time manipulate the polity parameters within which that action is rendered possible (Lawrence and Suddaby, 2006: 248).

Scholarship working on sustainability is aware of the importance of policy–polity relations – and this is true both for the more prescriptive work, as well as that focused on analysing the implementation of environmental policy and sustainability in practice. In prescriptive writing, for example, Lange *et al.* in their search for 'suitable governance for SD [sustainable development]' (2013: 404) start from a mapping of typologies of policy–polity–politics inter-relations in order to meet their objective to establish 'which governance mode (or mix of modes) is best suited to promoting SD and therefore ought to be advocated' (Lange *et al.*, 2013: 404). Asara *et al.* in their discussion on how to bring about the transformative de-growth agenda argue that sustainability-induced change takes place in a broader context of transformations involving 'multiple scales and system levels, from the local to the regional, national and international levels, and functional levels such as the markets, states and civil society' (2015: 379). Yet they do not propose how scholarship might grasp these processes to bring about a de-growth agenda.

In analytically focused work, scholars have both theorised the policy–polity relationship and demonstrated its analytical purchase for explaining causality. For example, Arts *et al.* in their analysis of environmental policy change have argued extensively that sustainability reforms

> are also expressions of political changes, for instance, a movement and expansion of politics, administration and policy beyond the current formal institutional frameworks.
>
> (2006: 95)

Focusing their analysis on dialectics between political modernisation and policy arrangements, they have demonstrated inter-relations between changing state, market and civil society, on the one hand, and environmental policy choices on the other. They have argued that local environmental reforms are also expressions of broader political changes which both explain choices made and reproduce political modernisation. This has led to a variety of policy arrangements and a diffusion of political power (Arts *et al.*, 2006: 103). Adger *et al.*, (2003) have argued that 'thick' analysis of environmental decision-making better enables research to grasp broader governing values of efficiency, effectiveness, equity and legitimacy and how they are reproduced in different settings, paying attention to interactions of institutions, context and scale.

Horlings and Marsden (2011, 2014), exploring sustainable rural development paradigms, have demonstrated the importance of analysing territorial politics in choices over 'green economy' forms, i.e. the bio-economy versus the eco-economy. The bio-economy, they contend, addresses 'environmental problems by technocratic and corporatist modes of policy-making' (Horlings and Marsden, 2011: 4) whereas the eco-economy contains 'a stronger institutional embeddedness of activities in specific contexts of space and place' (Horlings and Marsden, 2014: 7).[3] 'Territory' has also featured as an explanation of environmental policy choices, including of failures to implement, when clashes between 'intergovernmental and supranational parts of the EU' (Jordan and Tosun, 2013: 258) are analysed as primary causes of such failure.

As well as markets and territory, scholars have also examined institutions of science and knowledge as shaping policy–polity interactions in sustainability governance. For example, Hajer's seminal work on ecological modernisation highlighted the critical role of the social constructions of science and technology in the formation of discourse coalitions implementing different forms of ecological modernisation (Hajer, 2005 [1995]). More recently, Wesselink *et al.*, (2013) have shown how policy discourses – in which science and knowledge about nature play a critical role – both shape, and are shaped by, their socio-political environment, especially when they become dominant discourses. Research on science must therefore not only focus on the types of knowledge used in policy choices, but more specifically on 'how expertise is interwoven with other constitutive elements … in policy discourses'

(Wesselink *et al.*, 2013: 4). The challenge on how to integrate science and technology in the policy–polity analysis of sustainability is one also taken up by Toke (2011a, 2011b). Re-examining ecological modernisation's causal factors in the case of the wind energy industry, he has argued that grass roots social movements have been key actors in developing alternative technologies and discourses which have re-shaped both industrial practice and public policy choices.

States, markets and civil society, territory and knowledge have thus all been individually identified by scholars as political factors in the implementation of environmental and sustainability policy. Such factors are consequently considered contributory to processes of achieving sustainability. However, to date, these have not been brought together and theorised in a single analytical framework. This means that research is potentially missing important causes and interactions between sustainability, territory, regulation and knowledge. This is particularly so if we place an emphasis on actor strategies of engagement stressing actor politics of interdependence. In this vein, in an analysis of the transformation of Scottish fisheries implementing sustainable fisheries production, I observed that actors not only engaged in political work on interdependencies between natural and socio-ecosystems to effect change, but crucially on other forms of interdependence, namely: of territory, states and markets, and knowledge (Carter, 2014). Bringing these different phenomena together within a single analytical framework can therefore both expand the scope of the study and place actors and institutions at the heart of the analysis. This can also enable research to respond to certain demands made in the literature:

> [W]ho holds 'power' over the concept of sustainable development and how sustainable development is constructed need a framework which objectively sets out the range of available resources and identities for actors.
>
> (Springett, 2013: 79)

In reflecting on the contours and contents of such a framework, it is important not to treat on-going transformations in the wider polity as mere context or blocking exogenous structure. Rather, in line with political sociological applications of public policy analysis, political transformations must become a research object in their own right. This means that whereas a major emphasis of sustainability scholarship on the political system has been to consider how it must first change in order to bring about a particular meaning of sustainability, it is important to keep in mind that the political system is already changing, that broader changes have imperatives beyond environmental degradation, and that these potentially provide power resources and identities for actors shaping sustainability choices.

Expanding the perimeters of sustainability's interdependence

Following from these arguments, in this second section I expand upon the parameters of sustainability's interdependencies. More specifically, I argue that the transformation of the Westphalian system of state power, coupled with the stabilisation of a public/private co-regulation and the democratisation of knowledge use in policy-making, are collectively re-assigning political authority across the polity. This re-assignment of authority plays across and within three forms of interdependence: interdependence and territory; inter-dependence and regulation; and interdependence and knowledge. My aim in this section is to clarify actor tensions at the heart of these three interdependencies.

Interdependence and territory

The first interdependence is interdependence and territory, and more specifi-cally the politics of interdependence between territories. The latter is an old phenomenon whose meaning is constantly changing. Whereas the political modernisation trend for many hundreds of years consolidated 'political com-munity … institutions and policies … on a single spatial scale: that of the nation-state' (Jeffery and Wincott, 2010: 167), today the political meaning of 'territory' can neither be reduced to 'state sovereignty' (Sassen, 2013), nor can political interdependencies between territories be viewed as occurring between states constructed as single entities (Faure and Douillet, 2005; Faure *et al.*, 2007). Rather, the past few decades have witnessed an extensive re-territorialisation of politics through the re-scaling of political power and authority which has contributed to 'the denaturalisation of the nation state as a given and self-evident unit of politics' (Kauppi *et al.*, 2016: 2). This has put in motion a form of interdependence which is both *inter*-state and *infra*-state, altering the organisation and the meaning of state sovereignty along the way (MacCormick, 1993).

On the one hand, processes of regional integration, such as European inte-gration, have resulted in the construction of new territories (e.g. the Euro-pean Union). In Europe, responding to changing patterns in both regional and global markets and the opening up of the European regional trading area in the 1950s, new political arrangements were put in place to govern newly created markets. As has been well documented in an extensive literature on Europe and its Union (Holland, 1980; Nugent, 2010, Smith, 2010 [2004]; Wallace and Wallace, 2000), the pooling of state sovereignty in the early Treaties, for example, the European Economic Community Treaty (1957), not only gave powers to newly created EU public bodies (e.g. the Council of Ministers and the European Commission) to regulate negative market inte-gration through dismantling barriers to trade, but also powers for positive integration and the setting of common policies with redistributive effects to address new inequalities incurred (e.g. inequalities between regions, for example).

As is well known, since then and through myriad Treaty reforms and a deepening process of European integration, in which the authority to regulate has been assigned to EU bodies over an increasing number of policy areas, EU public bodies today exercise a wide range of legislative authority. This is especially the case for EU environmental policy which is considered one of the most advanced of EU public policies (Eckersley, 2004: 251; Jordan and Adelle, 2013).[4] These processes have thus over time created a new political supranational territory, whose authority is 'omnipresent, incomplete and de-politicised' (Jullien and Smith, 2014) and whose legitimacy is highly contested.[5]

On the other hand, within nation states themselves, a variety of processes of sub-state territorial politics, regional national identity projects and decentralisation and regionalisation agendas have additionally resulted in a reassignment of political power from central public actors to newly created regional and local bodies (Pasquier *et al.*, 2013). These have unfolded in many European countries, e.g. in France, Spain, Belgium, Italy, UK. Many of these regional governments and parliaments hold primary and secondary legislative powers over a range of policy areas and increasingly these competences are ones which are shared with both central state-wide administrations and EU bodies (for example, environmental policy). In short, the re-dispersal of political power both above and within the nation state has recreated new political communities and identities, new public actors and new public policy resources at different scales and hence a range of political territories with emerging (and potentially conflicting) territorial interests.

However, it is not only through constitutional and international treaty reform processes that processes of territorialisation can be observed. Within public policies and policy instruments, new frontiers of public action have been created which are not aligned with constitutionally-determined administrative jurisdictions and in which references to territory are important legitimating features (Carter and Smith, 2008). This has happened in particular with regard to territorial development projects, whose contents have not only provided redistributive policies correcting spatial disparities, but also political authority resources for local politicians to reinforce their visibility (Douillet, 2003). This has also been the case for EU public rules on geographical indications of products. These do not always map neatly onto the borders of administrative territories but, on the contrary, can result in the construction of new borders within a region, including creating a space for the empowering of new groups of actors controlling these new borders and new categories of territorial thought to justify them (e.g. Melton Mowbray pork pie production, Rumford, 2008).

In this territorialisation process, environmental policy has played a critical role (Salles, 2006). Interdependencies of territory have been shown to be critical in analyses of environmental responsibility (Salles, 2011), studies on how 'the biophysical' acts on governing choices (Fortin *et al.*, 2010; Le Floch, 2014), and on issues of rural and industrial patrimony (Cazals, 2012). Within fisheries and marine environmental governance, marine parks, marine regions and sea fisheries regions have all been delimited as newly defined governing regions existing alongside 'old' administrative territories, creating ever more complex territorial interdependencies (Van Hoof, 2015).

These new forms of territorial interdependence represent 'a major new – or renewed – challenge to the theory and practice of politics in Europe' (Jeffery and Wincott 2010: 168). A key such challenge is to theorise actor politics of this territorial interdependence. This, I contend, takes place at the interface of the interdependence and, more specifically, in actor struggles over the frontiers of territorial public action.

Actors become engaged in territorial frontier politics for several reasons. For example, increasing integration of economies and enhanced patterns of global trade mean that policy solutions for policy problems occurring in one territory frequently require to be negotiated with those of other territories. Geographical and environmental interconnections can also engender politics of territorial interdependence: for example, water pollution caused by factories in one territory can flow downstream into another territory encouraging a collective environmental responsibility (Salles, 2009). Spillover effects between one territory and another can also be regulatory in their origin in that choices made in the governing of a marine park to ban fishing might cause conflicts for regional managers governing fisheries who then mobilise to assert authority over marine park territory.

Whereas in many of these examples the politics of the interdependence turns on questions of effectiveness and legitimacy, there might also be a democratic dimension. This is because in some cases, the creation of a renewed territory and/or decentralisation processes might not only be concerned with the re-assignment of policy competence to regional actors, but also more fundamentally be about changing notions of democracy (Blatrix, 2012). For example, in France, processes of decentralisation have often been intertwined with processes of participatory local democracy (Blatrix, 2012); whereas in the UK, devolution in 1999 was premised upon institutionalising new norms of inclusive, participatory democracy in both Scotland and Wales (complementing representative democracy) and this was built into both the Scottish and Welsh Assemblies' rules of procedure (Carter, 2013a; Chaney *et al.*, 2007; Henderson, 2005). The question of which territory 'should' act can thus also amount to a question of which form of democracy should inform decision-making.

Several epistemological understandings are built into the theory of the politics of territorial interdependence in this book, and which are coherent with a political sociology of institutions. First, care will be taken not to overly emphasise formal divisions of policy competence as determinants of 'appropriateness' in scales of territorial action. Rather, it is understood that 'territory doesn't just happen; it has to be worked for' (Painter, 2010: 1105). What I am interested in is how the formal organisation of power can be navigated by actors through exploiting interdependencies: actors can engage in 'border-work' (Rumford, 2008), mobilising territory not just as a jurisdiction, but as a category of thought to exert their influence through multi-positioning strategies. Territory can be used as a power resource by actors to interpret their interests, to make them converge with those of others, to define problems and adopt policy instruments. Of course, this does not always happen: at times, territorialisation is more about spatial allocation of socio-economic resources than the reinforcement of local political authority (Douillet, 2003). Yet at other times, actors can and do evoke representations of territory in policy-making when (i) setting the frontiers of regulatory instruments, (ii) determining which actors are eligible to take decisions and at what scale, and (iii) legitimising compromises reached in situations of conflict or uncertainty (Carter and Smith 2008, 2009).

Local social relations are critical in these strategies (Itçaina, 2010). This is to keep to the fore the knowledge that there is contingency at the heart of the interdependence (Faure and Douillet, 2005). The relation between the 'local' and the 'global' is a complex one: it does not necessarily result in a powerlessness of local actors when responding to global pressures (Le Galès and Voelkow, 2003),[6] nor does it denote a triumph for the regional scale of political organisation. Regions and local territories are neither inevitable 'victims' of policy choices made by others in distant or far-away places and nor does their politics necessarily denote the end of other logics of action (Faure and Douillet, 2005).

Some scholars have labelled these processes of re-organisation of state power as a *de*-territorialisation of politics and, in so doing – as criticised by Morgan – they 'write territory out' of their analysis (2007: 1247). Accordingly, economic and political interconnections are deemed so pervasive that there is 'no definable regional territory to rule over' (Amin, 2004: 34, 36). This is not the position adopted here. Rather, the politics of territorial interdependence seeks to study both 'when' and 'how' territorial integrity and coherence is constructed (Itçaina, 2010) and made to join up – or not – with other territories operating at different scales. In other words, the politics at the frontier is not necessarily one of either keeping something out or moving together towards common policies at a higher scale. It can also be one of drawing something in (Ozga and Dubois-Sahik, 2015). This is when, for example, actors mobilise EU resources to solve local problems without any legal imperative to do so and in order to contribute to territorial continuity on a regional scale (Ozga and Dubois-Sahik, 2015; Pasquier, 2005).

Interdependence and regulation

The second interdependence is 'interdependence and regulation' and turns on the politics of interdependence between public and private regulation. Whereas interdependence between states and markets has been recognised for some time (Polanyi, 1957 [1944]), the meaning of this interdependence can and does change. States can and have intervened in markets in different ways through creating, manipulating and sanctioning behaviour both on the market and outside the market, through maintaining a market society (Le Galès and Scott, 2008: 307). Today, this intervention is framed within state projects of neoliberalisation (Goven and Pavone, 2015), 'bureaucratic revolution' (Le Galès and Scott, 2008), the advent of the regulatory state (Borraz, 2007), and public sector reform (Bevir and Rhodes, 2003). Importantly, within these reform projects, state action has not broken this interdependence, but has altered it:

> [W]e do not associate neoliberalism with a noninterventionist state but rather with a shift in the style and goals of intervention.
>
> (Goven and Pavone, 2015: 311)

> [S]tandards serve as an instrument of renegotiation of the state's role and influence in a changing society.
>
> (Borraz, 2007: 82)

Indeed, recent public sector reform and the rise of the regulatory state have opened up different possible avenues for states to continue to interact with markets. These include privatisation of public services, creation of new markets for public services, corporate management, and co-regulation and standardisation (Bevir and Rhodes, 2003; Boraz, 2007; Cafaggi, 2010; Le Galès and Scott, 2008).

Within these changing relationships, a significant development which is of importance here has often included the redistribution of power from public to private regulators and shifts in regulation from the state to civil society (Borraz, 2007; Cafaggi, 2010; Lascoumes and Le Galès, 2007). Understood as 're-regulation and not de-regulation' (Borraz, 2007: 59), this delegation of policy-making powers to economic actors (at times working with NGOs) has stabilised 'policy capture by economic interests' (Borraz, 2007: 59). This has led to the emergence of a proliferation of new forms of public/private co-regulation, voluntary private self-regulation, and incentive-based instrumentation operating at different scales: domestic, EU and global (Borraz, 2007; Cafaggi, 2010). More and more regulation is being set by private collective actors acting 'voluntarily' to self-regulate their own practices than by public administrations (in consultation with stakeholders) (Djama *et al.*, 2011; Fulponi 2006; Ponte *et al.*, 2011). Forms of co-regulation and private self-regulation include: third-party certification on a global scale (e.g. the Aquaculture Stewardship Council certification), standardisation (e.g. of food safety standards on an EU–wide scale), eco-labels, voluntary self-regulation, codes of best practice, standards for supermarket brands, business-to-business contracts, and private labels.

The causes of this proliferation of private regulation are multiple and depend also on the scale at which this regulation is being set. Some scholars link this trend to actors' responses to public sector reform within different governmental traditions (Bevir and Rhodes, 2003, 2010). Others locate the causes of new transnational private regulation in the 'weaknesses of states as global rule-makers' (Cafaggi 2010: 26) and the political mobilisation of international NGOs to 'govern' global trade, for example, on grounds of sustainability in the absence of international trade policy rules on this very issue (Bartley, 2007; Cashore, 2002). Others still analyse this trend as part and parcel of new regimes of governance, such as 'experimentalist governance' (Overdevest and Zeitlin, 2014). At a local scale, the origins of private self-regulation can be traced to politicisation strategies of e-NGOs in domestic markets, expressly seeking to create a responsibility for economic actors in long supply chains to self-regulate locally (Carter, 2015).

Of course, the rise of private regulation does not necessarily mark the end of public regulation. For example, although EU environmental policy is often presented as having sought to replace coercive traditional hierarchical regulation with flexible policy instruments, target-orientated legislation and voluntary agreements (Löber, 2011), nonetheless 'command and control' instruments still remain (e.g. in the EU common fisheries policy). Indeed, in environmental policy where moves towards the adoption of market-based instruments have been 'encouraged by environmental economists' (Halpern, 2010: 40; McAfee, 2012), EU environmental policy nevertheless remains characterised by its instruments, both public and private (Halpern, 2010). Indeed, according to Duit *et al.* (2016), the literature has paid too much attention to private regulation. In response, Duit *et al.* (2016) have re-affirmed the significance of the 'environmental state' and public regulation as a research object, criticising what they perceive as an assumption in the literature that 'the state' is no longer relevant in environmental public policy analysis.

In the examination of interdependence between public and private regulation set out in this book, I do not take a position on which regulation – public or private – 'should' be the focus of research. Rather, as with territory, the focus is on the politics at the heart of the interdependence. Actually, in debates on regulation, most scholarship supports the argument that public and private regulation are interdependent processes and that this relationship is a critical one to study (Auld, 2014; Gulbrandsen, 2014; Havinga, 2006; Rosenau, 2007; Rumpala, 2009):

> [T]he fact that state regulatory agencies do not exist in [a] vacuum but rather in a framework of mutually constitutive interaction with non-governmental institutions and actors, suggests that to focus on a single institution in whatever sector gives one a limited perspective on the regulatory process.
>
> (Grabosky, 1995: 202)

More fundamentally, political work at the interface between public and private instruments can explain governing transitions (Sergent, 2014). Frontier politics within this interdependence are on the boundary between the public and private and are over tensions and potential contradictions versus complementarities between the two domains. For example, public/private interdependencies have been at the core of analyses on environmental problem construction explaining how and why sectoral problems can become public or not (Candau *et al.*, 2012). In some cases, for example in standardisation, standards can complement public action and enable policy effectiveness; in others, standards are rather a substitute for public action building new markets through privately-regulated competitiveness (e.g. through product certification) (Borraz, 2007). As with interdependencies of territory, the politics of interdependence of public and private regulation thus raise issues of effectiveness and legitimacy (Loconto and Fouilleux, 2014). Further, they also raise questions of democracy. Democratic issues emerge because private regulation can entail a depoliticisation of public policy through a delegation of 'technical' decisions to 'experts' (Borraz, 2007). Choosing private regulation therefore also means choosing a particular form of democracy ('technical democracy'), potentially legitimised by technical expertise and the role of civil society in setting rules.

But the form of democracy is not the only democratic consideration at stake in managing frontiers between public and private regulation. The frontier politics of this interdependence can also affect democracy in terms of citizen engagement. As argued by Hay (2007), moving to the private realm has fundamental repercussions on politics. This is because the depoliticisation of politics brought about through the shifting of choices from the public to the private realm is not merely an issue of technicisation. More profoundly, it means removing politics from the realm of 'political deliberation' and placing it instead in the 'realm of fate' (Hay, 2007). The realm of fate is a paralysing place where citizens disavow their potential agency to determine their own destiny (Stoker, 2009: 85–86), increasing the sense for ordinary citizens that politics is 'beyond our influence' (Hay, 2007).

In this realm, the mechanism for engagement is the market, where citizens exist as consumers. For example, whereas in an imagined 'green state' citizens would be holders of constitutional rights, in the world of private regulation, citizens often participate as consumers, buying (or not) eco-labelled products: 'The message is be an active consumer not an active citizen' (Stoker, 2009: 85). As argued extensively by Muller (2010), this comes about not least because public policy has a democratic and societal role stabilising value choices and seeking to govern behaviour accordingly. It provides a public space for struggles over society's global vision and the future role for an industry therein. Consequently, the more a society opts for private instruments, the less the vision is a societal one and the more it becomes the vision of dominant economic actors and NGOs.

For all these reasons the frontier between the public and the private

> is not merely an issue of academic interest; it is itself the product of ongoing political and social struggles to define and delineate the appropriate reach of public … responsibility.
>
> (Hay 2010: 11)

In line with embedding this analysis within a political sociology of institutions, the main analytical attention is on how actors mediate the frontiers between public and private regulation and how they manage policy interactions. This process is not expected to be automatic or natural, but will depend on the 'political work' of actors (defined later in this chapter) which can be analysed to assess how boundaries are being drawn (Hay, 2010: 11) and their consequences for enactments of different sustainability narratives.

Interdependence and knowledge

The third interdependence concerns the politics of the relationship between different forms of knowledge used in policy-making and in policy implementation. The origins of this interdependence are manifold and can be traced back along (at least) four interweaving trajectories collectively contributing to the end of the 'era of knowledge use in the singular' (Demszky and Nassehi, 2012: 171).

First, critiques of the 'linear model of expertise' advanced within the literature on Science and Technology Studies (STS) and Sociology of Scientific Knowledge (SSK)[7] led to appeals both for greater reflexivity by scientists in the production of 'regulatory science' used for policy-making and for an opening out of organisations of technology and governing arrangements to new forms of public participation (Beck, 2011; Jasanoff 2003a; Miller and Edwards 2001; Waterton and Wynne 1996; Weingart 1999). As argued by Jasanoff: 'It was long assumed that the diffusion of fundamental knowledge into application was linear and unproblematic' (2003a: 228). Different problems resulting from the application of this philosophy of knowledge use observed by scholars led them to question this assumption. Rather than being unproblematic, the problems included *inter alia* alleged scientific fraud, policy failures, conflicts of scientific values with broader societal values and the inability of this model to address complex (so-called 'wicked') problems (Bavington, 2010; Beck, 2011; Fischer, 2000; Jasanoff, 2003a; Pielke, 2007). Consequently, society had to go

> far beyond the model of 'speaking truth to power' that once was thought to link knowledge to political action.
>
> (Jasanoff, 2003a: 225)

This led to calls for reform in understandings of accountability both within the scientific community and within decisional committees and for reflexive approaches to policy-making grasping recursive relationships between science and politics (Weingart, 1999; Wynne, 1996).

These changes were interwoven with a second trajectory on new policy-making cultures brought about within broader transformations of the state, markets and civil society and processes of political modernisation within Europe. Participatory governance was endorsed as a core component of this change (*Arts et al.*, 2006; Turnhout and Van der Zouwen, 2010). Referred to by some as a 'qualitative turn in political practice' (Demszky and Nassehi, 2012: 173), in a variety of policy areas and territorial contexts increased participation and engagement of 'lay people' and non-experts in public decision-making was observed (Blatrix, 2002, 2009; Dodge, 2009; Madsen and Noe, 2012; Wesselink *et al.*, 2013). This change in governance practice signified a political response to critical social theory critiques of expert decision-making (by Habermas, 1962, 1973) and the call for participatory democracy to complement the limits of representative democracy (Blatrix, 2002, 2009). For some scholars, the use of non-traditional versus traditional knowledge indicated a change in approach towards deliberative governance:

> Involving experience based knowledge in the course of policy is in many aspects a kind of revolution against the usual political order ... [it] not only relativizes, but questions the authoritative knowledge of experts and policy makers.
>
> (Demszky and Nassehi, 2012: 172)

For others, societal shits towards participatory governance were considered to be motivated by more instrumental reasons within reforms of the regulatory state (see the previous discussions on interdependence and regulation).

In these interpretations, what was at stake were knowledge gaps: groups in civil society held the knowledge and expertise necessary for effective governance and which was not (always) known by public administrations. This was observed when participatory democracy was translated into practice whereby local knowledge was framed as critical for the effective implementation of local decisions (Blatrix, 2012) and domestic knowledge was viewed as essential for the effective development of EU policies (Douillet and de Maillard, 2010). As with interdependence and territory, research has shown that viewing knowledge use as either a protest or as a complement to the existing order does not provide either/or explanations; rather, it reveals tensions at the heart of the knowledge interdependence. For example, the deliberative imperative is not necessarily in opposition to representative democracy (Blatrix, 2002); and public actors request new types of expert knowledge not only for instrumental reasons, but also for symbolic ones (Douillet and de Maillard, 2010). Alternative knowledge use therefore can have different consequences for governing modes, ranging from change to adaptation at the margins (Connelly, *et al.*, 2006; Salles *et al.*, 2014). Accordingly, the meaning of participation is found through examining the practice.

These discussions have been further interwoven with a third trajectory linked to calls from within the scientific community for greater humility and reflexivity.[8] Within these debates, a knowledge utilisation literature and practice began to stabilise, setting recommendations for improving science policy dialogue and identifying new and innovative ways to fulfil governmental demands for usable science and bring about reflexive approaches to knowledge exchange

(Dilling and Lemos, 2011; Jordan and Russel, 2014; Nutley *et al.*, 2007; Young *et al.*, 2014). This has encouraged discussions both on questions of policy impact and on the governance of the science–policy interface as entailing recursive relationships (Görg *et al.*, 2016).

Whereas the first three of these trajectories indicate change towards the interdependence of knowledge forms driven mainly by elites, a fourth trajectory can be identified which is rather change brought about within civil society itself. This is witnessed on the one hand by the increase in the number of sites of knowledge production (e.g. the growth in the number of consultants in new knowledge markets) (Jasanoff, 2003a). On the other hand, knowledge has been used politically by groups of actors as an institutional resource of empowerment. There are myriad examples of both militant and politically engaged public action led by local groups, stakeholders or citizen associations who have mobilised as 'holders of knowledge' to gain entry to decisional arenas from which they had been hitherto excluded.

This has been the case, for example, for certain types of 'citizen science' through the creation of citizen observatories of nature generating new data (Salles *et al.*, 2014); within the strategies of environmental NGOs desiring a seat at the decisional table and who have self-represented as holders of new kinds of expertise to gain entry to government (e.g.: Weisbein, 2015 on Surfrider Foundation Europe; Ollitrault, 2001 on French e-NGOs); producers who have mobilised as holders of data over different scales and timeframes than those used in scientific surveys to access policy-making and challenge the contents of policy instruments governing their production practices (e.g. Carter, 2013b on Scottish fishers); and stakeholders who have mobilised to produce knowledge in collective social arrangements (e.g. Hage *et al.*, 2010).

When making these arguments, effective and legitimate policy is often represented by these civil society actors as ultimately dependent upon a 'broader definition of what counts as evidence' (Elgert, 2010: 386). This can lead to strategies of 'institutional positioning' (Douillet and de Maillard, 2010), both by stakeholders and inside organisations. Given that the type of knowledge being used will determine the nature of the problem being identified and its possible solution, we can also include in this fourth trajectory powering within the scientific community itself to destabilise dominant scientific theories and scientists (Roger, 2010). What is considered the 'best available scientific expertise' emerges from internal struggles within disciplines to bring about a hierarchy of knowledge (Itçaina *et al.*, 2016; Roger, 2010). When scientists define the boundaries of 'their' discipline and its authority for policy-making they simultaneously determine hierarchies of power amongst scientists, in some cases constructing new fields of research associated with shifts within industries (Itçaina *et al.*, 2016; Roger, 2010).

Whereas the trajectories described above are global ones, in the sense that they can be observed in many policy domains, they have been particularly noticeable within environmental policy (Blatrix, 2012; Bouleau *et al.*, 2009;

Deuffic, 2012). Indeed, one of the premises of 'sustainable development' as set out in the Brundtland report (see Chapter 2) is that the difficult decisions which need to be taken in its name should have 'the support and involvement of an informed public and of NGOs [as well as] the scientific community, and industry' (WCED 1987: 36). To give effect to this norm, participatory democracy and public consultation have been important governing techniques used for integrating environmental concerns into a range of planning choices (Blatrix, 2012). Furthermore, the nature of current environmental challenges for managing biodiversity and adaptation to climate change pose problems which are complex and over which science is uncertain, hence further encouraging 'compound knowledge use' in problem solving (Boezeman, 2015; Jordan and Russel, 2014).

For all these reasons, multiple knowledge forms can and have been 'used' in policy-making. These include, *inter alia*, regulatory science, applied science, expertise, technical expertise, user knowledge, everyday knowledge, traditional knowledge, citizen science and local knowledge. In the present analysis, the focus is on the politics of the interdependence between these knowledge types and actor struggles over knowledge use. Crucially, however, as with the other interdependencies to be examined, the intention is not to set criteria to hierarchise particular forms of knowledge use, but rather to study how actors themselves decide which knowledge is appropriate and why.

In stating this, we note that this has already become a critical debate within STS. This is because, in response to the production and uptake of multiple knowledge forms and/or the involvement of civil society in technological decisions, oppositions can be observed amongst STS scholars on how to study the frontier politics of this interdependence (in STS vocabulary, the 'boundary work': Gieryn, 1995). In their well-known (and controversial) paper on the Third Wave of science studies, Collins and Evans (2002) launched a discussion on what they perceived as a philosophical problem resulting from the opening up of the domain of technological decision-making to increased public participation. This 'intellectual problem', as they described it, turned on the question of

> the value of scientists' and technologists' knowledge and experience as compared with others' knowledge and experience.
>
> (Collins and Evans, 2002: 236)

Less than a decade later, they described the problem they perceived in this way:

> Demands for increased public participation in science, however, have the tendency to lead to a 'levelling of the epistemological playing field' and to a collapse of the concept of expertise.
>
> (Collins *et al.*, 2010: 186)

This related to two further problems as they saw it: the 'problem of exten-sion', and specifically how far participation should be extended without col-lapsing the category of expertise; and the problem of deconstruction, namely that whereas SSK had deconstructed expertise, it had failed to reconstruct it to define the difference between 'good expertise' and 'bad-pseudo expertise' (Collins *et al.*, 2010: 192). If scientists no longer had special access to the truth, they questioned, why should their advice be especially valued (Collins and Evans, 2002: 236)?[9]

To be clear, this essentialist approach to the analysis of the politics of inter-dependence and knowledge is not the one built into the analytical framework being developed here. Rather, in keeping with new political sociological scholarship on this very question (e.g. Frickel and Moore, 2006),[10] the focus is on how actors in interaction with institutions draw the frontier and deter-mine what counts and what does not count as legitimate knowledge in policy implementation. Nonetheless, it is important to note that this position could very well be encountered in the research. This is to understand that there are power relations at the frontier between different knowledge use in decision-making; specifically, that

> the institutionalisation of new knowledge derives from the production of new power relations, and that the institutionalisation of new power rela-tions leads to new modalities and fields of knowledge.
>
> (Cibele *et al.*, 2010: 51)

Finally, as will have become apparent, the politics of interdependence and knowledge forms takes place within contested debates over knowledge use in policy-making, which raise issues of effectiveness, legitimacy and democracy. The democratisation of knowledge use in policy-making is action which is far from settled and remains highly controversial in both scientific debates and in practical applications. Frontier politics is to be anticipated – not only within science (and within disciplines) over questions of environmental impact (as discussed in Chapter 2) – but also between political usages of different knowledge types. It is expected that these 'politicisations'[11] will not only be led by elites, but also by groups within civil society seeking to politi-cise industrial practices through self-representing as holders of knowledge to gain access to political spaces of governance.

Conceptual tools for studying the politics of sustainability interdependence

So far in this chapter, I have argued that incorporating policy–polity dialectics in the analysis of the governing of sustainability in European fish farming aquaculture requires the development of a new analytical framework which can both recompose sustainability and integrate it within a wider theory of change. This has required extending the parameters of the research object through a discussion of on-going changing regularities of territory, regulation and knowledge interdependencies in European democratic states.

In this final section, drawing on combined scientific insights from a political sociology of politics, institutions and economic governance (Itçaina *et al.*, 2016; Jullien and Smith, 2014; Kauppi, 2010; Kauppi, *et al.*, 2016; Lawrence and Suddaby, 2006; Merand, 2008; Muller, 2015; Roger, 2010), I seek to complete the construction of my analytical framework through a focus on concepts and methods. This requires integrating the sustainability politics of nature–society interdependencies, on the one hand, with the frontier politics of interdependence and territory, regulation and knowledge on the other – and defining them collectively as a research object of sustainability interdependence. To bring this about, I argue, requires first defining 'sustainability' heuristically as an institution (public action principle) whose social meaning is achieved through actor conflicts over sustainability narratives; and, second theorising sustainability's on-going construction via the concept of institutional and political work.

We already know that political sociological approaches have been pivotal in making the case for studying industries as critical sites for grasping public action and politics (Hall, 1986; Jobert and Muller, 1987; Jullien and Smith 2014), including sustainability politics (Carter *et al.*, 2014; Sergent, 2014). These approaches understand that industries are brought into relation with society through the institutions which govern them and the 'social, political, contextual, historical construction of economic activity' (Jullien and Smith, 2014: 4). Studying sectoral choices enables us to learn about wider changes in society. In these approaches, industries are viewed as more than supply and demand considerations (by comparison with neo-classical or economic deterministic accounts of industries). Their central governing institutions are understood to extend beyond social, financial and employment rules (the main focus of the Varieties of Capitalism or 'VoC' literature)[12] (Jullien and Smith, 2014).

More specifically, an industry (in this case, European fish farming aquaculture) is conceptualised as being governed through four categories of institutionalised relationship (IRs): production, commercial, employment and financial (Jullien and Smith, 2014). From this perspective, the governance of an industry includes the setting and implementation of rules and norms on production, including on sources of raw materials and/or farming considerations (e.g. fish feed contents for farmed fish; stocking densities of fish in cages; fish farm site location); on commercial issues, including on product types or markets (e.g. setting standards for organic salmon); on employment conditions (e.g. on training farm managers with new feed technology); and on finance (e.g. banking loans; farm site start-up funds). Authors have demonstrated that dominant social meanings of an industry and the interests of its respective governors are the result of the interplay between actors and these rules and norms, so that the sum total of this governing activity contributes to the very definition of an industry at any given time and in any given place (Carter and Cazals, 2014; Gorry *et al.*, 2014; Itçaina *et al.*, 2014; Jullien *et al.*, 2014).

Governing choices regulating the production, commercial, employment and financial IRs of an industry are not deemed 'spontaneous' or 'self-evident', but are socially constructed through this governing activity and are contingent (it could have been otherwise) (Jullien and Smith, 2014). Structure is built into the analysis through attention to institutionalised logics and routines which are understood to stabilise industrial identities (at times becoming entrenched) through the production and reproduction of dominant institutions and actors (Itçaina *et al.*, 2016; Sergent, 2014).

At the core of this analysis lies the concept of 'political work' (Jullien and Smith, 2014) or 'institutional work' (Lawrence and Suddaby, 2006). This concept find echoes with those discussed above within analyses of interdependencies, e.g. border work (Rumford, 2008). This concept emerges from an underlying epistemological understanding that change does not 'just happen'; nor can it 'be taken for granted' (Lawrence and Suddaby, 2006: 217). Change, incremental or otherwise, always results from the political and institutional work of public and collective private actors – and this is also the case for continuity, which is not void of agency[13] but is worked for.

Political work involves 'behaviour that both discursively and interactively seeks to change or reproduce institutions by mobilizing values' (Jullien and Smith, 2011: 14). It has been shown to include collective institutional practices of problematisation (defining industrial conditions as public problems for government), instrumentation (setting rules and norms) and legitimation (justifying choices made) (Smith, 2013; see also, Gilbert and Henry, 2012; Itçaina *et al.*, 2016; Paul, 2007; Roger, 2010). Scholars have also observed strategies of politicisation aimed at opening phenomena up to political controversy and contestation (Kauppi *et al.*, 2016) and rendering value choices explicit (Smith, 2013). This is to grasp

> politics as an activity, it is then something contingent – it is always possible to act otherwise, even if the results of the alternatives may be the same. Within this horizon we can define 'politicisation' as an active use of contingency, of rendering something contested or controversial.
>
> (Kauppi, *et al.*, 2016: 3)

Political activity can also depoliticise phenomena, closing them down to controversy and contestation, for example, through strategies of technicisation, when actors seek to hide value choices at stake within discussions on technical expertise (Smith, 2013). Political and institutional work can therefore involve a range of actor practices, all of which need studying. These include, *inter alia*: the framing of industrial conditions as public problems; the making of arguments; the building of political alliances; the building of new organisations or the changing of strategies of engagement in pre-existing organisations; the setting of instruments; and strategies of politicisation, mediatisation or technicisation of conflicts. Depending on the political situation, this work can be to 'create' new institutions (institutionalisation), 'maintain' institutions already in place (re-institutionalisation) or 'disrupt' already existing institutions (de-institutionalisation) (Jullien and Smith, 2014; Lawrence and Suddaby, 2006). The concept of political work thus denotes types of political engagement which can (and will) be studied empirically.

Following from these discussions, and staying within political sociological approaches to industry governing, in my case studies I treat sustainability heuristically as a core institution (public action principle) governing European fish farming aquaculture (impacting on actor choices in all four IRs of the industry). Defining sustainability in this way enables me to capture it as a 'socially embedded [process] … what [actors governing aquaculture] do' (Cleaver and Franks, 2005: 3). Defining sustainability as an institution does not mean, however, according it an essentialist meaning. Rather, and as the work carried out in Chapter 2 has revealed, different alternative narratives on nature–society interdependencies exist which can be mobilised by actors to render sustainability meaningful to them.

To grasp how actors have mobilised around these respective narratives, I mobilise the concepts of political and institutional work. In particular, I trace political and institutional work over long periods of time and examine how its accumulation has allowed for changing sustainability policy trajectories. I examine how sustainability's meaning has been created; how it has been disrupted; and how it has been maintained or changed in the face of opposition. I study how certain 'settlements of conflicts' (Bartley, 2007) of sustainability are brought about, whereby dominant actors govern in a situation of tension and alternatives (Schneiberg, 2007), and when and how this institution exercises its authority (Bevir, 2011; François, 2011).

This permits me to generate new knowledge on how actors in this industry 'fashion, … identify … and render … actionable their conceptions of interests' (Hay, 2007: 117). Core research questions generated are: 'how is "sustainability interdependence" being institutionalised?'; 'which arguments, instruments and actors dominate and why?'; and 'what are the alternatives'?[14] Finally, these contests are expected to be emotional (Mouffe, 2000, 2005) constructing industrial identities (Merand, 2008), whereby some actors and arguments will be excluded, given that a 'fully inclusive rational consensus' is neither possible nor at issue (Mouffe, 2000: 6).

In this analysis, two different kinds of political struggle are anticipated to interact, which lie at the heart of the politics of sustainability interdependence. The first is over policy directions and competing definitions of nature–society interdependencies (i.e. between (i) sustainability as depoliticised ecology–economy interdependencies; (ii) sustainability as politicised ecology–economy–society interdependencies; (iii) sustainability as ecosystem approaches problematising dynamic ecology–society interdependencies; and (iv) sustainability as de-growth re-ordering ecology–society–economy hierarchies). Here, the work carried out in Chapter 2 to disentangle the different narratives of sustainability will help me to study these actor struggles in practice.

As was argued, sustainability is a contested concept. This presents a challenge for how to capture it empirically, given that many actors may act in the name of 'sustainability'. In Chapter 2, I therefore organised each narrative around its core components which can provide key points of difference to be mobilised to categorise my empirical material. Addressing this challenge, I examine actor struggles first in relation to the various components which constitute the interdependence at the heart of each narrative, only seeking in a second 'movement' to categorise political work in relation to the narratives.[15] Therefore, even in situations where actors do not openly discuss their action using the term 'sustainability', if in their institutional work they have depoliticised ecology–economy interdependencies, then I consider them to be mobilising around a particular interpretation of sustainability. To help me in this analysis, I have additionally already used the terms 'depoliticise' and 'politicise' in my description of the interdependence.

Importantly, in proceeding thus, I do not propose to limit my analysis of political work on sustainability to how actors mobilise over interdependencies between nature and society. Rather, as I have already argued, the change in research object to sustainability interdependence means that I treat interdependence and territory, regulation and knowledge as *part and parcel* of the politics of sustainability and hence as part of the research object. This is to hypothesise that when governing key problems facing this industry, ecological, economic and social issues will not only be brought into relation with one another (classic sustainable development), but also into relation with these broader interdependencies unfolding within the political system (sustainability interdependence). To be clear, my research question is not how do territory, regulation and knowledge intervene on sustainability choices. Rather, the question is how do the *frontier politics of interdependence* affect the institutionalisation of sustainability? This is to place a focus on actors and their strategies of engagement. More precisely, do actors work at the interface of territory, regulation or knowledge interdependencies to create/maintain/disrupt sustainability? Does the way actors handle the politics of interdependencies cause stalemate or provide the seeds for change?

The second struggle to be grasped is of a different order. This is over the very authority to govern and enact particular nature–society interdependencies in the first place, i.e. the authority to influence (Genieys and Hassenteufel, 2012; Genieys and Smyrl, 2008). This struggle is critical because society's sectoral enactment of the global (Muller, 2010; 2015) not only concerns how a society enacts the contents of global change, but also how power is redistributed in this process. This is to define politics as 'action which seeks to change the distribution of power' (Frickel and Moore, 2006: 10).

Research carried out applying a politics of industry approach asking 'who governs industries' has shown that not only public elites and producers govern industries through mobilisation of their respective resources, but that potentially all actors along the supply chain can govern an industry. Governors can include: providers of raw materials (e.g. in fish farming, this would include fish feed manufacturers), processors, retailers (e.g. supermarkets), NGOs, social movements and scientists (Carter and Cazals, 2014). Institutional work can therefore also be mobilised as a concept to measure power in governing practice:

> [I]nstitutional work requires[s] resources, which are available to some actors and not others. A critical view of institutional work could begin to examine how those resources are distributed and controlled, and by whom.
>
> (Lawrence and Suddaby, 2006: 247)

Critically, it is my hypothesis that the frontier politics of interdependence and territory, regulation and knowledge also potentially offer resources and

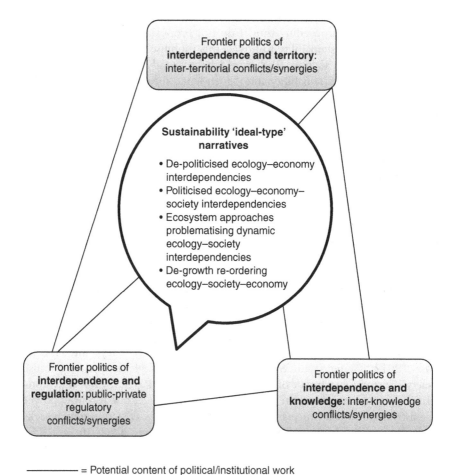

——————— = Potential content of political/institutional work

Figure 3.1 Dynamic institutionalist framework capturing political tensions of 'sustainability interdependence'.

identities for actors to alter power relations. How actors mobilise their frontier politics either allows for entry of a range of actors into decision-making, or it closes this down.

As outlined in the introductory chapter, this gives rise to two sets of research questions:

i   How does the politics of interdependencies of territory, regulation and knowledge ultimately come to shape the way in which sustainability problems are rendered governable? For example, in their 'institutional/political work' do actors mobilise around the frontier politics of these interdependencies to create contingency and open space for alternative policy directions in the governing of sustainability? Or do they instead experience any contradictions contained therein as constraints?

ii How does the institutionalisation of sustainability as a governing problem for European fish farming aquaculture in turn affect broader processes of transformation of the political system? How does the institutional and political work of actors in European fish farming aquaculture contribute to the perpetuation of these societal interdependencies?

In summary, through mobilising concepts of institution and institutional work, I expect to be able to demonstrate how actors in the governing of European fish farming aquaculture have over time not only mobilised over different interdependencies between nature and society, but also over the interdependence and territory, regulation and knowledge. Operating at a meso level of analysis and providing synthetic accounts of changing trajectories of governing practice, I aim not only to provide new knowledge on how the political system has shaped aquaculture, but also to explain how, in turn, aquaculture has shaped the political system.

Conclusions

Through defining sustainability as an institution and integrating it into a wider theory of political change, this chapter has reframed this book's research object around sustainability interdependence. As I argued at the start of this chapter, this shift is consistent with lessons from public policy analysis of policy–polity dialectics. The chapter makes two contributions: first, by combining interdependence and territory, regulation and knowledge together into a single framework for studying the implementation of sustainability, and second, by theorising these interdependencies from within a political sociology of politics of industry governing. This provides key concepts of institution and institutional and political work. It is envisaged that the application of these concepts to different sectors and territories in European fish farming aquaculture will enable me to compare the institutionalisation of different policy–polity directions of sustainability interdependence governing this industry. A strong working hypothesis is that through the political work of

Table 3.1 Research guidance and checklist

	ITD Terr + Nature/ Society	ITD Reg + Nature/ Society	ITD Know + Nature/ Society	Sustainability Interdependence
Social constructions of sustainability	1 De-politicised ecology–economy interdependencies? 2 Politicised ecology–economy–society interdependencies? 3 Ecosystem approaches problematising dynamic ecology–society interdependencies? 4 De-growth re-ordering ecology–society–economy hierarchies?			**Description of policy problems and governing problems** Which policy problems can be identified? Which governing problems can be identified?
Re-assignments of political power and authority; Resources and identities	Which territories? Which frontier politics of territory?	Which public–private regulation? Which frontier politics of regulation?	Which knowledge types? Which frontier politics of knowledge?	
Scale of political intervention	Administrative scale	Sectoral scale and public–private scale	Socio-ecological scale	
Actor struggles over sustainability	**Evidence of political and institutional work over:** Economy–ecology win–win? Economy–ecology–social justice? Socio-ecosystem? Ecology–social justice? **Political and institutional work underpinned by actor arguments on:** Nature as technological fix? Nature as resources? Nature as benefits? Nature as wilderness? Nature as vulnerable? Nature as attractive? Intergenerational equity? Historical fidelity (natural, industrial, cultural heritage)? Social contract? Precautionary approach? Property rights? Relation between fish farming and natural/biophysical processes (water, fish, other species)? Relation between fish farming and landscapes? Relation between fish farming and markets; growth? Relation between fish farming and community; employment?			**Description of sustainability interdependence** **Policy substance and tensions** Tensions between sustainability narratives regulating problems? Mitigation/ compensation strategies? **Which assignments of political authority?** Dominant actors and institutions? Alternatives: shadow actors; emergent institutions and marginal knowledges? **Tensions between scales?** Do scales of political intervention overlap or mismatch? **Forms of democracy?** Utilitarian, individualist, efficiency? Distributional, communitarian (state/ local scale)? Equity? Representative? Participatory? Technical? Active or consumer citizens? Realm of ownership? Realm of fate?
Actor struggles over wider ITDs	**Evidence of political and institutional work over:** Frontiers of territorial legitimacy? Synergies? Discordance? New identities? New resources? Jurisdictions? Territorial categories of thought? Territory as a political authority and/or spatial allocation of socio-economic resources? Eligibility of actors?	**Evidence of political and institutional work over:** Frontiers of co-regulation? Synergies? Discordance? New identities? New resources? New regulations/ instruments? Policy capture by economic interests? By NGOs? Private standards as complementary or substitution to public action? Eligibility of actors?	**Evidence of political and institutional work over:** Frontiers of different knowledge types? Synergies? Discordance? New identities? New resources? New knowledges? Institutional positioning? Science as independent truth of 'how things are'? Knowledge as partial, contested and fragile? Investment in new research? Eligibility of actors?	

actors, sustainability-induced change will become entangled with these broader transformations to the extent that they all become co-constituted. This in turn will create hybridisation of sustainability narratives in policy practice whereby transversal interdependencies of territory, regulation and knowledge emerge as important vehicles for their entanglement.

In the next three chapters of the book, I will apply these research questions to investigate the politics of sustainability interdependence of European fish farming aquaculture in different sectors and territories compared and around different industry challenges. I will compare which nature–society interdependencies have been institutionalised over time and through which frontier politics of territory, regulation and knowledge interdependencies. Through comparing I will be able to grasp differences and specificities in public action, and potentially common patterns of institutionalisation. I will mobilise both the narratives described in Chapter 2 and the three interdependencies described in this Chapter 3 to generate criteria to describe different political settlements of sustainability interdependence in the three sectors (salmon, trout, seabass/seabream) and territories. As I have explained in the introductory chapter, throughout, my aim is to go 'inside' governing practices not only to document the practices observed, but also to describe how actors themselves represent their political and institutional work when at the interface of interdependence.

Finally, there is a wider argument on democracy to be teased out throughout what follows. This is because how actors manage the respective interdependencies at stake will not only affect the sustainability of European fish farming aquaculture and its environmental impact, but also how societal democracy is being re-institutionalised. As was argued in Chapter 2, different interpretations of sustainability have also incorporated different theories of politics, notably the utilitarian versus distributional. Similarly, the politics of each interdependence and territory, regulation and knowledge has a democratic dimension. Consequently, the settlement of different forms of sustainability interdependence in each sector and territory will also include a democratic settlement whose causes and consequences will be raised in each empirical chapter and elaborated upon in the general conclusions to this book in Chapter 7.

Notes

1 With some notable exceptions, e.g. see Eckersley, 2004.
2 This is by no means to suggest that all work on 'sustainability *and* governance' is mobilised from a sociological and constructivist perspective. On the contrary, there are many works which embrace the overlap between 'governance for' and 'governance and' sustainability. This is especially the case when scholars who hold a specific prescriptive definition of sustainability proceed to study whether this version of sustainability has been adopted (or not) in actor practice in different settings (e.g. see Lafferty, 2004; Mol, 2000). As argued by York and Rosa (2003), often the focus of this research is on whether the governing instruments which have been proposed to achieve sustainability have been put in place, rather than some other measure of change.
3 In the study carried out by Horlings and Marsden (2014), examples given of eco-economy place-based innovative products included: saffron production in Tuscany; Gregoriano cheese, local bio-energy heating systems in Finland; energy production based on biomass of forests in Valtellina in Lombardia, Italy.
4 EU instruments on the environment are wide-ranging, covering air and water pollution, waste management, biodiversity and protection of wildlife, the countryside and the marine environment.
5 Most recently evidenced by Brexit of course!

6 Even within exclusive EU common policies, such as fisheries, regional public actors can and do mobilise to design their own policy instruments locally and in line with regional conceptions of sustainability justice (see Carter, 2014).

7 The 'linear model of expertise assumed that policy makers first defined the political problem, then requested and received expert advice, and finally a political decision was taken' (Weingart, 1999). The model assumed a separation of 'the political' from 'the science' – a separation which has been extensively refuted by scholarship within STS. See also, Beck, 2011; Hoppe, 2005; Pielke, 2007.

8 These calls from the scientific community have been encouraged in some cases by regulatory conditions for accessing public research funding.

9 There is no space here to go into the details of this debate, but see, e.g. Collins and Evans, 2002; Jasanoff, 2003b; Wynne, 2003. For similar debates, see Merton, 1942 and Gieryn, 1995.

10 This approach invites the following research questions: 'Which knowledge gets reproduced? Who gains access to that knowledge? What kinds of knowledge are left "undone"?' (Frickel and Moore, 2006: 8).

11 I define politicisation in the section below.

12 In the VoC approach, the focus is on 'national' institutions as dominant (and uniform) ones providing opportunities or constraints for firms' strategic leeway in product innovation related to activities within the employment institutionalised relationship (IR) (i.e. works councils; collective agreements) and, to a lesser extent, the finance IR (Lange, 2009). This is to overlook political struggles around interdependencies of territory, regulation and knowledge within production and commercial IRs where significant institutions can arguably be found operating at multiple scales – e.g. regional and European.

13 As the impression is sometimes given by deterministic applications of historical institutionalism.

14 The question of alternatives is essential to critical analysis as it enables the researcher to problematise politics, rather than just assuming that politics is something which is well known; see Kauppi, *et al.* (2016).

15 Here I draw on Bevir and Rhodes' discussion cautioning against the reification of traditions (2003: 140).

References

Adger, W., Brown, K., Fairbrass, J., Jordan, A., Paavola, J., Rosendo, S., Seyfang, G. 2003. 'Governance for sustainability: towards a "thick" analysis of environmental decision making', *Environment and Planning*, A(35): 1095–1110.

Amin, A. 2004. 'Regions unbound: towards a new politics of place', *Geografiska Annaler*, 86 B(1): 33–44.

Arts, B., Leroy, P., Tatenhove, J. 2006. 'Political modernisation and policy arrangements: a framework for understanding environmental policy change', *Public Organization Review*, 6: 93–106.

Asara, V., Otero, I., Demaria, F., Corbera, E. 2015. 'Socially sustainable degrowth as a social–ecological transformation: repoliticizing sustainability', *Sustainability Science*, 10: 375–384.

Auld, G. 2014. 'Confronting trade-offs and interactive effects in the choice of policy focus: specialized versus comprehensive private governance', *Regulation & Governance*, 8: 126–148.

Bartley, T. 2007. 'Institutional emergence in an era of globalization: the rise of transnational private regulation of labour and environmental conditions', *American Journal of Sociology*, 113(2): 297–351.

Bavington, D. 2010. *Managed Annihilation: An Unnatural History of the Newfoundland Cod Collapse*, Vancouver: UBC Press.

Beck, S. 2011. 'Moving beyond the linear model of expertise? IPCC and the test of adaptation', *Regional Environmental Change*, 11(2): 297–306.

Behagel, J., Turnhout, E. 2011. 'Democratic legitimacy in the implementation of the Water Framework Directive in the Netherlands: towards participatory and deliberative norms?', *Journal of Environmental Policy & Planning*, 13(3): 297–316.

Bevir, M. 2011. 'Public administration as storytelling', *Public Administration* 89(1): 183–195.

Bevir, M., Rhodes, R. 2003. *Interpreting British Governance*, Routledge: London.

Bevir, M., Rhodes. R. 2010. *The State as Cultural Practice*, Oxford: Oxford University Press.

Blatrix, C. 2002. 'Devoir débattre. Les effets de l'institutionnalisation de la participation sur les formes de l'action collective', *Politix*, 15(57): 79–102.

Blatrix, C. 2009. 'La démocratie participative en représentation', *Sociétés Contemporaines* 2(74): 97–119.

Blatrix, C. 2012. 'Des sciences de la participation: paysage participatif et marché des biens savants en France', *Quaderni* [online]: 79.

Boezeman, D. 2015. *Transforming Adaptation: Authoritative Knowledge for Climate Change Governance*, Enschede, The Netherlands: Ipskamp Drukkers.

Borraz, O. 2007. 'Governing standards: the rise of standardization: processes in France and in the EU', *Governance*, 20(1): 57–84.

Bouleau, G., Argillier, C., Souchon, Y., Barthélémy, C., Babut, M. 2009. 'How ecological indicators construction reveals social changes – the case of lakes and rivers in France', *Ecological Indicators*, 9(6): 1198–1205.

Cafaggi, F. 2010. 'New foundations of transnational private regulation'. EUI Working Paper, RSCAS: 53.

Candau, J., Deldrève, V., Deuffic, P. 2012. 'Publicisation controlée de problèmes territoriaux autour de l'eau: le cas des Pertuis charentais (France)', *SociologieS*, [online]: 1–14.

Carter, C. 2013a. 'Rethinking UK parliamentary adaptation in EU affairs: devolution and Europeanisation', *Journal of Legislative Studies*, 19(3): 392–409.

Carter, C. 2013b. 'Constructing sustainability in EU fisheries: re-drawing the boundary between science and politics?', *Environmental Science and Policy*, 30: 26–35.

Carter, C. 2014. 'The transformation of Scottish fisheries: sustainable interdependence from "net to plate"', *Marine Policy*, 44: 131–138.

Carter, C. 2015. 'Who governs Europe? Public versus private regulation of sustainability of fish feeds', *Journal of European Integration*, 37(3): 335–352.

Carter, C., Cazals, C. 2014. 'The EU's government of aquaculture: completeness unwanted'. In Jullien, B. and Smith, A. (eds) *The EU's Government of Industries: Markets, Institutions and Politics*, Abingdon: Routledge, pp. 84–114.

Carter, C., Smith, A. 2008. 'Revitalizing public policy approaches to the EU: "territorial institutionalism", fisheries and wine', *Journal of European Public Policy*, 15(2): 263–281.

Carter, C., Smith, A. 2009. 'What has Scottish devolution changed? Sectors, territory and polity-building', *British Politics*, 4(3): 315–340.

Carter, C., Cazals, C., Hildermeier, J., Michel, L., Villareal, A. 2014. 'Sustainable development policy: "Competitiveness" in all but name'. In Jullien, B. and Smith,

A. (eds) *The EU's Government of Industries: Markets, Institutions and Politics*, Abingdon: Routledge, pp. 165–189.

Cashore, B. 2002. 'Legitimacy and the privatization of environmental governance: how non–state market–driven (NSMD) governance systems gain rule-making authority', *Governance*, 15(4): 503–529.

Cazals, C. 2012. 'Examining the conventions of voluntary environmental approaches in French agriculture', *Cambridge Journal of Economics*, 36(5): 1163–1181.

Chaney, P., Mackay, F., McAllister, L. 2007. *Women, Politics and Constitutional Change: The First Years of the National Assembly for Wales*. Cardiff: University of Wales Press.

Christen, M., Schmidt, S. 2012. 'A formal framework for conceptions of sustainability – a theoretical contribution to the discourse in sustainability', *Sustainable Development*, 20: 400–410.

Cibele, C., Barroso, J., Carvalho, M. 2010. *Knowledge and Public Action. Sex Education in School*, Knowledge and Policy Report. Available at: www.knowandpol.eu/IMG/pdf/portugal_education_-_o2__ap2_-_corrige.pdf.

Cleaver, F., Franks, T. 2005. 'How institutions elude design: river basin management and sustainable livelihoods'. Bradford Centre for International Development, Research Paper no. 12. Available at: http://core.kmi.open.ac.uk/display/5659.

Collins; H., Evans, R. 2002. 'The Third Wave of science studies: studies of expertise and experience', *Social Studies of Science*, 32(2): 235–296.

Collins, H., Weinel, M., Evans, R. 2010. 'The politics and policy of the Third Wave: new technologies and society', *Critical Policy Studies*, 4(2): 185–201.

Connelly, S., Richardson, T., Miles, T. 2006. 'Situated legitimacy: deliberative arenas and the new rural governance', *Journal of Rural Studies*, 22(3): 267–277.

Demszky, A., Nassehi, A. 2010. 'Perpetual loss and gain: translation, estrangement and cyclical recurrence of experience based knowledges in public action', *Policy and Society*, 31(2): 169–181.

Deuffic, P., 2012. *Produire et discuter des normes environnementales: écologues et forestiers face à la biodiversité associée au bois mort*, Ecole Doctorale Sociétés, Politique, Santé publique (SP2), Bordeaux.

Dilling, L., Lemos, M.C. 2011. 'Creating usable science: opportunities and constraints for climate knowledge use and their implications for science policy', *Global Environmental Change*, 21(2): 680–689.

Djama, M., Fouilleux, E., Vagneron, I. 2011. 'Standard-setting, certifying and benchmarking: a governmentality approach to sustainable standards in the agro-food sector'. In Ponte, S., Gibbon, P. and Vestergaard, J. (eds) *Governing through Standards: Origins, Drivers and Limitations*, Basingstoke: Palgrave Macmillan, pp. 184–209.

Dodge, J. 2009. 'Environmental justice and deliberative democracy: how social change organizations respond to power in the deliberative system', *Policy and Society*, 28: 225–239.

Douillet, A-C. 2003. 'Les élus ruraux face à la territorialisation de l'action publique', *Revue française de science politique*, 53: 583–606.

Douillet, A-C., de Maillard, J. 2010. 'Les comités d'experts: une ressource institutionnelle pas toujours maîtrisée. le cas de la DG justice, liberté, sécurité', *Politique européenne* 3(32): 77–98.

Duit, A., Feindt, P. Meadowcroft, J. 2016. 'Greening Leviathan: the rise of the environmental state?', *Environmental Politics*, 25(1): 1–23.

Eckersley, R. 2004. *The Green State: Rethinking Democracy and Sovereignty*, Cambridge MA: MIT Press.

Elgert, L. 2010. 'Politicizing sustainable development: the co-production of globalized evidence-based policy', *Critical Policy Studies*, 3(3-4): 375–390.

Farrell, K., Kemp, R., Hinterberger, F., Rammel, C., Ziegler, R. 2005. 'From "for" to governance for sustainable development in Europe', *International Journal of Sustainable Development*, 8(1/2): 127–150.

Faure, A., Douillet, A-C. 2005. *L'action Publique et la Question Territoriale*, Grenoble: Presses Universitaires de Grenoble.

Faure, A., Leresche, J-P., Muller, P., Nahrath, S. (eds). 2007. *Action Publique et Changements d'Echelles: Les Nouvelles Focales du Politique*, Paris: L'Harmattan.

Fischer, F. 2000. *Citizens, Experts and the Environment: The Politics of Local Knowledge*, Durham and London: Duke University Press.

Fortin, M.J., Devanne, A.S., Le Floch, S. 2010. Le paysage politique pour territorialiser l'action publique et les projets de développement: le cas de l'éolien au Québec. *Développement Durable et Territoires* [online], 1(2) (September 2010), published online 23 September 2010. Available at: http://developpementdurable.revues.org/index8540.html.

François, P. (ed.). 2011. *Vie et Mort des Institutions Marchandes*, Paris: Presses de Sciences Po.

Frickel, S., Moore, K. (eds). 2006. *The New Political Sociology of Science Institutions, Networks, and Power*, Madison: The University of Wisconsin Press.

Fulponi, L. 2006. 'Private voluntary standards in the food system: the perspective of major food retailers in OECD countries', *Food Policy*, 31: 1–13.

Genieys, W., Hassenteufel, P. 2012. 'Qui gouverne les politiques publiques? Par-delà la sociologie des élites', *Gouvernement et Action Publique*, 2(2): 89–115.

Genieys, W., Smyrl, M. 2008. *Elites, Ideas and the Evolution of Public Policy*, London and New York: Palgrave.

Gieryn, T. 1995. 'Boundaries of science'. In Jasanoff, S., Markle, G., Petersen, J. and Pinch, T. (eds) *The Handbook of Science and Technology Studies*, London: Sage.

Gilbert, C., Henry, E. 2012. 'Defining social problems: tensions between discreet compromise and publicity', *Revue française de sociologie* (English), 1(53): 31–54.

Görg, C., Wittmer, H., Carter, C., Turnhout, E., Vandewalle, M., Schindler, S., Livorell, B., Lux, A. 2016. 'Governance options for science–policy interfaces on biodiversity and ecosystem services: comparing a network vs. a platform approach', *Biodiversity and Conservation*, 25: 1235–1252.

Gorry, P., Montalban, A., Smith, A. 2014 'The EU's government of pharmaceuticals: incompleteness embraced'. In Jullien, B. and Smith, A. (eds) *The EU's Government of Industries: Markets, Institutions and Politics*, Abingdon: Routledge, pp. 115–137.

Goven, J., Pavone, V. 2015. 'The bioeconomy as political project: a Polanyian analysis', *Science, Technology, and Human Values*, 40(3): 302–337.

Grabosky, P.N. 1995. 'Governing at a distance: self-regulating green markets'. In Eckersley, R. (ed.) *Markets, the State, and the Environment: Towards Integration*, South Melbourne, Australia: Macmillan Education, pp. 197–228.

Gulbrandsen, L. 2014. 'Dynamic governance interactions: evolutionary effects of state responses to non-state certification programs', *Regulation & Governance*, 8: 74–92.

Habermas, J. 1962. *The Structural Transformation of the Public Sphere: An Inquiry into a Category of Bourgeois Society*, Cambridge MA: MIT Press.

Habermas, J. 1973. *Theory and Practice*, Boston: Beacon Press.

Hage, M., Leroy, P., Petersen, A. 2010. 'Stakeholder participation in environmental knowledge production', *Futures*, 42: 254–264.

Hajer, M. 2003. 'Policy without polity? Policy analysis and the institutional void', *Policy Sciences*, 36: 175–195.

Hajer, M. 2005 [1995]. *The Politics of Environmental Discourse: Ecological Modernization and the Policy Process*, 2nd edition, Oxford: Oxford University Press.

Hall, P.A. 1986. *Governing the Economy: The Politics of State Intervention in Britain and France*, Oxford: Oxford University Press.

Halpern, C. 2010. 'Governing despite its instruments? Instrumentation in EU environmental policy', *West European Politics*, 33(1): 39–57.

Havinga, T. 2006. 'Private regulation of food safety by supermarkets', *Law and Policy* 28(4): 15–533.

Hay, C. 2007. *Why We Hate Politics*, Cambridge: Polity Press.

Hay, C. 2010. 'Introduction: political science in an age of acknowledged interdependence'. In Hay, C. (ed.) *New Directions in Political Science: Responding to the Challenges of an Interdependent World*, Basingstoke: Palgrave, pp. 1–24.

Henderson, A. 2005. 'Forging a new political culture: plenary behaviour in the Scottish Parliament', *Journal of Legislative Studies*, 11(2): 275–301.

Holland, S. 1980. *Uncommon Market: Capital, Class, and Power in the European Community*, New York: St. Martin's Press.

Hoppe, R. 2005. 'Rethinking the science–policy nexus: from knowledge utilization and science technology studies to types of boundary arrangements', *Poiesis & Praxis: International Journal of Ethics of Science and Technology Assessment*, 3(3): 199–215.

Horlings, L., Marsden, T. 2011. 'Towards the real green revolution? Exploring the conceptual dimensions of a new ecological modernisation of agriculture that could "feed the world"', *Global Environmental Change*, 21(2): 441–452.

Horlings, L., Marsden, T. 2014. 'Exploring the "new rural paradigm" in Europe: eco-economic strategies as a counterforce to the global competitiveness agenda', *European Urban and Rural Studies*, 21(1): 4–20.

Itçaina, X. 2010. 'Les régimes territoriaux de l'économie sociale et solidaire: le cas du Pays Basque français', *Géographie, économie, société*, 12(1): 71–87.

Itçaina, X., Roger, A., Smith, A. 2014. 'The EU's government of wine: switching towards completeness'. In Jullien, B. and Smith, A. (eds) *The EU's Government of Industries: Markets, Institutions and Politics*, Abingdon: Routledge, pp. 35–56.

Itçaina, X., Roger, A., Smith, A. 2016. *Varietals of Capitalism: A Political Economy of the Changing Wine Industry*, Ithaca and London: Cornell University Press.

Jasanoff, S. 2003a. 'Technologies of humility: citizen participation in governing science', *Minerva*, 41(3): 223–244.

Jasanoff, S. 2003b. 'Breaking the waves in science studies: comment on H.M. Collins and Robert Evans, "The Third Wave of science studies"', *Social Studies of Science*, 33(3): 389–400.

Jeffery, C., Wincott, D. 2010. 'The challenge of territorial politics: beyond methodological nationalism'. In Hay, C. (ed.) *New Directions in Political Science: Responding to the Challenges of an Interdependent World*, Basingstoke: Palgrave, pp. 167–188.

Jennings, P., Zandbergen, P. 1995. 'Ecologically sustainable organizations: an institutional approach', *The Academy of Management Review*, 20(4): 1015–1052.

Jobert, B., Muller, M. 1987. *L'Etat en action*, Paris: PUF.

Jordan, A., Adelle, C. (eds). 2013. *Environmental Policy in the EU: Actors, Institutions and Processes*, 3rd edition, Abingdon: Routledge.

Jordan, A., Russel, D. 2014 'Embedding the concept of ecosystem services? The utilisation of ecological knowledge in different policy venues', *Environment and Planning C: Government and Policy*, 32: 192–207.

Jordan, A., Tosun, J. 2013. 'Policy implementation'. In Jordan, A., Adelle, C. (eds) *Environmental Policy in the EU: Actors, Institutions and Processes*, Abingdon: Routledge, pp. 247–266.

Jullien, B., Smith, A. (eds). 2014. *The EU's Government of Industries: Markets, Institutions and Politics*, Abingdon: Routledge.

Jullien, B., Pardi, T., Ramirez-Perez. 2014. 'The EU's government of automobiles: from harmonization to deep incompleteness'. In Jullien, B. and Smith, A. (eds) *The EU's Government of Industries: Markets, Institutions and Politics*, Abingdon: Routledge, pp. 57–83.

Kauppi, N. 2010. 'The political ontology of European integration', *Comparative European Politics*, 8(1): 19–36.

Kauppi, N., Palonen, K., Wiesner, C. 2016. 'Controversy in the garden of concepts: rethinking the "politicisation" of the EU'. Joint Working Paper Series of Mainz Papers on International and European Politics (MPIEP) No. 11 and Jean Monnet Centre of Excellence 'EU in Global Dialogue' (CEDI) Working Paper Series No. 3. Mainz: Chair of International Relations, Johannes Gutenberg University.

Lafferty, W. (ed.). 2004. *Governance for Sustainable Development*, Cheltenham: Edward Elgar Publishing.

Lagroye, J. 1991. *Sociologie Politique*, Paris: Presses de la FNSP.

Lange, K. 2009. 'Institutional embeddedness and the strategic leeway of actors: the case of the German therapeutical biotech industry', *Socio-Economic Review*, 7: 181–207.

Lange, P., Driessen, P., Sauer, A., Bornemann, B., Burger, P. 2013. 'Governing towards sustainability-conceptualizing modes of governance', *Journal of Environmental Policy & Planning*, 15(3): 403–425.

Lascoumes, P., Le Galès, P. 2007. 'Introduction: understanding public policy through its instruments. From the nature of instruments to the sociology of public policy instrumentation', *Governance*, 20(1): 1–21.

Lawrence, T., Suddaby, R. 2006. 'Institutions and institutional work'. In Hardy, C., Lawrence, T. and Nord, W. (eds) *Sage Handbook of Organization Studies*, London: Sage, pp. 215–254.

Le Floch, S. 2014. 'Les bords de Garonne et leurs nouveaux riverains', *Ethnologie française*, 1(44): 165–172.

Le Galès, P., Scott, A. 2008. 'Une révolution bureaucratique britannique? Autonomie sans contrôle ou 'freer markets, more rules', *Revue française de sociologie*, 49(2): 301–330.

Le Galès, P., Voelkow, H. 2003, 'Introduction: the governance of local economics'. In Crouch, C., Le Galès, P., Trigilia, C. and Voelkow, H. (eds) *Local Production Systems in Europe. Rise or Demise?* Oxford: Oxford University Press, pp. 1–24.

Löber, S. 2011. 'The place of knowledge in policy-making processes: an assessment of three EU environmental policy instruments'. In Atkinson, R, Terizakis, G. and Zimmermann, K. (eds) *Sustainability in European Environmental Policy: Challenges of Governance and Knowledge*, London: Routledge, pp. 77–93.

Loconto, A., Fouilleux, E. 2014. 'Politics of private regulation: ISEAL and the shaping of transnational sustainability governance', *Regulation & Governance*, 8(2): 166–185.

MacCormick, N. 1993. 'Beyond the sovereign state', *Modern Law Review*, 56(1): 1–18.

McAfee, K. 2012. 'The contradictory logic of global ecosystem services markets', *Development and Change*, 43(1): 105–131.

Madsen, M., Noe, E. 2012 'Communities of practice in participatory approaches to environmental regulation. Prerequisites for implementation of environmental knowledge in agricultural context', *Environmental Science & Policy*, 18: 25–33.

Merand, F. 2008. 'Les institutionnalistes (Américains) devraient-ils lire les sociologues (Français)?', *Politique européenne*, 2(25): 23–51.

Merton, R.K. 1942. 'Science and technology in a democratic order', *Journal of Legal and Political Sociology*, 115–126.

Miller, C., Edwards, P. 2001. 'Introduction: the globalization of climate science and climate politics'. In Miller, C. and Edwards, P. (eds) *Changing the Atmosphere: Expert Knowledge and Environmental Governance*, Cambridge MA: MIT Press, pp. 1–30.

Mol, A. 2000. 'The environmental movement in an era of ecological modernisation', *Geoforum*, 31: 45-56.

Morgan, K. 2007. 'The polycentric state: new spaces of empowerment and engagement?', *Regional Studies*, 41(9): 1237–1251.

Mouffe, C. 2000. *Deliberative Democracy or Agonistic Pluralism*. Wien: Reihe Politikwissenschaft/Institut für höhere Studien, Abt. Politikwissenschaft 72. Available at: http://nbn-resolving.de/urn:nbn:de:0168-ssoar-246548.

Mouffe, C. 2005. *On the Political*, London: Routledge.

Muller, P. 2010. 'Les changements d'échelles des politiques agricoles: introduction'. In Hervieu, B., Mayer, N., Muller, P., Purseigle, F. and Rémy, J. (eds), *Les Mondes Agricoles en Politique*, Paris: Presses de Sciences Po, pp. 339–350.

Muller, P. 2015. *La société de l'efficacité globale. Comment les sociétés se pensent et agissent sur elles-mêmes*, Paris: PUF.

Nugent, N. 2010. *The Government and Politics of the European Union*, Basingstoke: Palgrave Macmillan.

Nutley, S.M., Walter, I., Davies, H. 2007. *Using Evidence: How Research Can Inform Public Services*, Bristol: Policy Press.

Ollitrault, S. 2001. 'Les écologistes français, des experts en action', *Revue française de science politique* 51(1): 105–130.

Ollitrault, S. 2004. 'Des plantes et des hommes de la défense de la biodiversité à l'altermondialisme', *Revue française de science politique*, 54(3): 443–463.

Overdevest, C., Zeitlin, J. 2014. 'Assembling an experimentalist regime: transnational governance interactions in the forest sector', *Regulation & Governance*, 8: 22–48.

Ozga, J., Dubois-Sahik, F. 2015 'Referencing Europe: usages of Europe in national identity projects'. In Carter, C. and Lawn, M. (eds) *Governing Europe's Spaces: European Union Re-Imagined*, Manchester: Manchester University Press, pp. 111–129.

Painter, J. 2010. 'Rethinking territory', *Antipode*, 42: 1090–1118.

Pasquier, R. 2005 ' "Cognitive Europeanization" and the territorial effects of multilevel policy transfer: local development in French and Spanish regions', *Regional & Federal Studies*, 15(3): 295–310.

Pasquier, R., Simoulin, V., Weisbein. J. 2013. *La gouvernance territoriale: pratiques, discours et théories*. Paris: l'Extenso.

Paul, K. 2007. 'Food for thought: change and continuity in German food safety policy', *Critical Policy Analysis*, 1(1): 18–41.

Pielke, R.A. Jr. 2007. *The Honest Broker. Making Sense of Science in Policy and Politics*, Cambridge: Cambridge University Press.

Polanyi, K. 1957 [1944]. *The Great Transformation: The Political and Economic Origins of Our Time*, Boston, MA: Beacon Press.

Ponte, S., Gibbon, P., Vestergaard, J. (eds). 2011. *Governing through Standards: Origins, Drivers and Limitations* Basingstoke: Palgrave Macmillan.

Revel, R., Blatrix, C., Blondiaux, L., Fourniau, J-M., Hériard Dubreuil, B., Lefebvre, R. 2007 'Introduction'. In Revel, R., Blatrix, C., Blondiaux, L., Fourniau, J-M., Hériard Dubreuil, B. and Lefebvre, R. (eds) *Le débat public: une expérience française de démocratie participative*, Paris: Éditions La Découverte, pp. 9–34.

Richardson, J. (ed.). 1996. *European Union: Power and Policy-making*, London: Routledge.

Roger, A. 2010. 'Constructions savantes et légitimation des politiques européennes', *Revue française de science politique*, 60(6): 1091–1113.

Rosenau, J. 2007. 'Governing the ungovernable: the challenge of a global disaggregation of authority', *Regulation & Governance* 1: 88–97.

Rumford, C. 2008. 'Introduction: citizens and borderwork in Europe', *Space and Polity*, 12(1): 1–12.

Rumpala, Y. 2009. 'La "consommation durable" comme nouvelle phase d'une gouvernementalisation de la consommation', *Revue française de science politique*, 5(59): 967–996.

Salles, D. 2006. *Les défis de l'environnement, démocratie et efficacité*, Paris: Editions Syllepses.

Salles, D. 2009. 'Environnement: la gouvernance par la responsabilité?', *VertigO – la revue électronique en sciences de l'environnement* [online], hors série 6.

Salles, D. 2011. 'Responsibility based environmental governance', *S.A.P.I.E.N.S* [online], 4(1): 1–7.

Salles, D., Bouet, B., Larsen, M., Sautour, B. 2014. 'A chacun ses sciences participatives. Les conditions d'un observatoire participatif de la biodiversité sur le Bassin d'Arcachon', *ESSACHESS. Journal for Communication Studies*, 7(1/13): 93–106.

Sassen, S. 2013. 'When territory deborders territoriality. Territory, politics', *Governance*, 1(1): 21–45.

Schneiberg, M. 2007. 'What's on the path? Path dependence, organizational diversity and the problem of institutional change in the US economy, 1900–1950', *Socio-economic Review* 5(1): 47–80.

Sergent, A. 2014. 'Sector-based political analysis of energy transition: green shift in the forest policy regime in France', *Energy Policy*, 73: 491–500.

Smith, A. 2010 [2004]. *Le gouvernement de l'Union européenne. Une sociologie politique*, 2nd edition, Paris: LGDJ.

Smith, A. 2013. 'Policy-making within the European Commission: problematization, instrumentation and legitimation', *Journal of European Integration*, 36, 55–72.

Springett, D. 2013. 'Editorial: critical perspectives on sustainable development', *Sustainable Development*, 21: 73–82.

Stoker, G. 2009. 'What's wrong with our political culture and what, if anything, can we do to improve it? Reflections on Colin Hay's "Why We Hate Politics"', *British Politics*, 4: 83–91.

Toke, D. 2011a. *Ecological Modernization and Renewable Energy*. Basingstoke: Palgrave.

Toke, D. 2011b. 'Ecological modernisation, social movements and renewable energy', *Environmental Politics*, 20(1): 60–77.

Turnhout, E., Van der Zouwen, M. 2010. '"Governance without governance": how nature policy was democratized in the Netherlands', *Critical Policy Studies*, 4(4): 344–361.

Van Hoof, L. 2015. 'Fisheries management, the ecosystem approach, regionalisation and elephants in the room', *Marine Policy*, 60: 20–26.

Van Tatenhove, J. 2013. 'How to turn the tide: developing legitimate marine governance arrangements at the level of the regional seas', *Ocean & Coastal Management*, 71, 296–304.

Wallace, H. 2000. 'Analysing and explaining policies'. In Wallace, H. and Wallace, W. (eds), *Policy-Making in the European Union*, 4th edition, Oxford: Oxford University Press.

Wallace, H., Wallace, W. (eds) 2000. *Policy-making in the European Union*, 4th edition, Oxford: Oxford University Press.

Waterton, C., Wynne, B. 1996. 'Building the European Union: science and the cultural dimensions of environmental policy', *Journal of European Public Policy*, 3(3): 421–440.

WCED (World Commission on Environment and Development). 1987. *Our Common Future* (Brundtland Report), Oxford: Oxford University Press.

Webb, C. 1977. 'Introduction: variations on a theoretical theme'. In Wallace, H., Wallace, W. and Webb, C. (eds) *Policy-making in the European Community*, 1st edition, London: J. Wiley.

Weingart, P. 1999. 'Scientific expertise and political accountability: paradoxes of science in politics', *Science and Public Policy*, 26(3): 151–161.

Weisbein, J. 2015. 'Capter et (co)produire des savoirs sous contraintes: le tournant expert de Surfrider Foundation Europe', *Politix*, 3: 93–117.

Wesselink, A., Buchanan, K., Georgiadou, Y., Turnhout, E. 2013. 'Technical knowledge, discursive spaces and politics at the science–policy interface', *Environmental Science and Policy*, 30: 1–9.

Wynne, B. 1996. 'Misunderstood misunderstandings: social identities and public uptake of science'. In Irwin, A. and Wynne, B. (eds) *Misunderstanding Science? The Public Reconstruction of Science and Technology*, Cambridge: Cambridge University Press, pp. 19–46.

Wynne, B. 2003. 'Seasick on the Third Wave? Subverting the hegemony of propositionalism: response to Collins and Evans (2002)', *Social Studies of Science*, 33(3): 401–417.

York, R., Rosa, E. 2003. 'Key challenges to ecological modernization theory', *Organization and Environment*, 16(3): 273–288.

Young, J.C., Waylen, K.A., Sarkki, S., Albon, S., Bainbridge, I., Balian, E., Davidson, J., Edwards, D., Fairley, R., Margerison, C., McCracken, D., Owen, R., Quine, C., Stewart-Roper, C., Thompson, D., Tinch, R., Van den Hove, S., Watt, A. 2014. 'Improving the science–policy dialogue to meet the challenges of biodiversity conservation: having conversations rather than talking at one-another', *Biodiversity and Conservation*, 23: 387–404.

Part II

Institutionalising sustainability in fish farming aquaculture

The place for European fish farming aquaculture as a potential response to a number of critical challenges confronting European societies is not at all certain. To grasp the complexity of any response made thus far necessitates empirical research to understand how actors have already been institutionalising sustainability as a governing institution in this industry. To organise such an empirical investigation, in Part I of this book, I set out a new analytical framework grasping policy–polity interdependencies including a conceptual diagram and methodological table detailing research questions to be asked and answered when analysing empirical material (Chapters 2 and 3). In this second part of the book, I present my empirical findings and compare the institutionalisation of sustainability within the governing of Scottish salmon, Aquitaine trout and Greek seabream and seabass. My approach throughout has been to grasp 'sustainability interdependence' as coming about through four 'tangled' processes of institutionalisation – nature/society; inter-territory; inter-public/private regulation; and inter-knowledge. From there I have developed conclusions regarding the institutionalisation of sustainability, and hence the sustainability interdependence of each territory and sector examined.

As explained in the introductory chapter, my approach to case-study design has been problem-orientated. Each chapter in Part II examines the institutionalisation of sustainability in European fish farming aquaculture around a central issue facing the industry: fish farm/environment interactions (Chapter 4), access to farm sites (Chapter 5), and sustainability as a food governance problem (Chapter 6). These are issues which emerged as critical ones during the course of EU-wide stakeholder consultation on the elaboration of the European strategy on sustainable development for aquaculture 2009 (Carter and Cazals, 2014: Carter et al., 2014); in my documentary analysis; and in the course of the interviews with actors in Scotland, Aquitaine, Greece (and Brussels). They correspond to three different scales of sustainability problems posed by aquaculture as identified by Soto et al. (2008: iv): farm scale; water body/watershed scale; and global market-trade scale.

Chapter 4 (fish farm/environment interactions) covers sustainability issues which can potentially be cast by actors as problems from the farm scale and up to the water body/watershed scale; Chapter 5 (access to farm sites) covers sustainability issues potentially perceived as problems extending from the water body/watershed scale and up to the community or regional scale; Chapter 6 (sustainability as a food governance problem) covers sustainability issues as potential problems stretching from the local market scale up to the global market/trade scale. Crucially, this is not to suggest that these are the scales at which each problem is actually governed.[1] As will be shown, this is a different question and one which will be addressed throughout what follows.

Note

1 Nor should they be confused with an administrative level of governance, as in the concept of multi-level governance, which denotes the formal jurisdiction of an administration and not necessarily the policy frontiers governing a problem.

References

Carter, C., Cazals, C. 2014. 'The EU's government of aquaculture: Completeness unwanted'. In Jullien, B. and Smith, A. (eds) *The EU's Government of Industries: Markets, Institutions and Politics*, Abingdon: Routledge, pp. 84–114.

Carter, C., Cazals, C., Hildermeier, J., Michel, L., Villareal, A. 2014. 'Sustainable development policy: "Competitiveness" in all but name'. In Jullien, B. and Smith, A. (eds) *The EU's Government of Industries: Markets, Institutions and Politics*, Abingdon: Routledge, pp. 165–189.

Soto, D., Aguilar-Manjarrez, J., Brugère, C., Angel, D., Bailey, C., Black, K., Edwards, P., Costa-Pierce, B., Chopin, T., Deudero, S., Freeman, S., Hambrey, J., Hishamunda, N., Knowler, D., Silvert, W., Marba, N., Mathe, S., Norambuena, R., Simard, F., Tett, P., Troell, M., Wainberg, A. 2008. Applying an ecosystem-based approach to aquaculture: principles, scales, and some management measures. In Soto, D., Aguilar-Manjarrez, J. and Hishamunda, N. (eds) *Building an Ecosystem Approach to Aquaculture*. FAO/Universitat de les Illes Balears Experts Workshop. 7–11 May 2007, Palma de Mallorca, Spain. FAO Fisheries and Aquaculture Proceedings No. 14, Rome: Food and Agriculture Organisation (FAO), pp. 15–35.

4   Sustainability interdependence and fish farm/environment interactions

Governing fish farming's environmental impact

Introduction

This chapter focuses on the sustainability governing of interactions between fish farms and the environment in Scottish salmon, Aquitaine trout and Greek seabass and seabream. As I set out in the introduction to this book, intensive and semi-intensive (Pillay, 1997) farming practices in each location have their origins in the 1970s and 1980s when small farm businesses began to apply newly emerging biophysical technologies to grow fish. During these early beginnings, fish farming was considered promising as a new industry which could not only make a profit, but, critically, could reduce pressure on wild fish stocks through substituting wild caught fish with farmed fish in the market place.

As the industry grew, however, it quickly became apparent that fish farming had its own environmental impact. Consequently, farming practices were increasingly brought under scientific and regulatory scrutiny. Yet, given knowledge gaps on environmental data, the tendency was to develop 'flexible approaches' to impact assessment which permitted development to take place (Pillay, 1997: 11). Indeed, the intensification of production and the setting up of more and more farms, together with the emergence of larger (and, in some cases, multinational) companies coincided with an increasing number of scientific articles (especially in ecology) documenting negative environmental impacts linked to production practices. Very early on this caused scholars to raise sustainability questions[1] about fish farming aquaculture:

> How large can the economy become relative to the constraints of the ecosystem?
>
> (Folke and Kautsky, 1989: 241)

In this chapter, I first briefly set out what the issue of fish farm/environment interactions 'could be' as discussed in this scientific literature on fish farming aquaculture.[2] I then proceed to examine actual governing practices of fish farm/environmental interactions and focus on how initial constructions of the problem in the 1980s and 1990s have more recently been worked upon by a range of actors around emerging controversies. Applying my institutionalist sustainability interdependence framework, in each case I will show how actor political struggles over nature–society interdependencies combined with struggles over the frontier politics of interdependence and territory, regulation, and/or knowledge have determined governing choices leading to particular 'settlements' of sustainability interdependence. These settlements are more or less stabilised depending on the sector and territory concerned and are frequently undercut with tensions which will be identified. Finally, I conclude by comparing the different social constructions and institutionalisations of fish farm/environment interactions between Scottish salmon, Aquitaine trout and Greek seabass and seabream, which turn both on environmental impacts of farms and on interactions between production and biophysical processes.

Throughout, the line of argumentation advanced is that there is nothing inevitable about these governing processes nor about the forms of sustainability interdependence that have been institutionalised. It is important to underline this point, because just as was the case for the literature on sustainability, in the literature on aquaculture articles have been published which treat the governance of fish farm/environment interactions from a normative perspective. For example, Frankic and Herschner (2003) describe what sustainable aquaculture *is* in the case of fish farm/environment interactions and list management measures and policy tools which *should* be put in place to govern aquaculture sustainably in accordance with their definition.

Whilst I recognise that this kind of approach intentionally politicises the lack of inevitability of negative environmental impacts of fish farms – arguing that impact is in part affected by the kinds of management measures in place – it is not the approach which I am adopting here. Rather than taking a checklist of sustainability governing tools and seeking to discover whether these have been implemented or not in the different case study sites, the overall aim is to compare how actors themselves have defined sustainability in fish farm/environment interactions and rendered it governable.[3]

The potential environmental impacts of fish farming aquaculture

According to Bailey (2008), already in the 1980s ecological concerns about aquaculture began to attract attention on a global scale (initially over effects of shrimp farming on mangroves). Since then, a range of potential negative environmental impacts of fish farms has been identified through scientific research, company data and government monitoring of farm sites. As I outlined in the introductory chapter, fish farming takes place in open water: the freshwater loch (salmon), seawater lochs or lagoons (salmon, seabass, seabream), and river site raceways (rainbow trout). Marine fish farming (salmon, seabream, seabass) takes place in net cages attached to floating structures in the water; freshwater fish farming (rainbow trout) by contrast takes place in river site raceways whereby part of the river water flow is diverted through the raceway and then discharged back into the river. In both cases, fish are raised, fed and treated in an open aquatic environment. For example, feeding takes place either manually or automatically with feeding machines (sometimes, in the case of marine farming, controlled by staff working in barges next to the farm site) through the scattering of fish feed pellets on the water above the cages or in the trout tanks. Similarly, although today the use of antibiotics has been more or less stopped in EU fish farming practice (see below) and replaced with vaccination of fish, medical treatments (for certain diseases and for parasites) take place whereby chemicals are introduced directly into the cages (and hence the water).

Discharges from farming practices can therefore have impacts on the marine and river environment. Fish farming can contribute to organic enrichment of underlying sediments of the sea (river) bed and the surrounding water body through organic discharges and the dispersion of organic wastes, e.g. uneaten and unabsorbed fish feed and faeces, either floating to the bottom of the sea or back into the river (Belias *et al.*, 2003; Folke and Kautsky, 1989; Frankic and Hershner 2003; Mantzaurakos *et al.*, 2007; Read and Fernandes, 2003; Russell *et al.*, 2011). Organic enrichment can have localised effects close to the cages (Keeley *et al.*, 2013; Russell *et al.*, 2011) and can affect the local benthic community – e.g. local bottom fauna and plankton communities (Mantzaurakos *et al.*, 2007) including causing local habitat loss (Wiber *et al.*, 2012), whereby local sediment can become anoxic (Folke and Kautsky, 1989). Fish farm sediment accumulation can create anoxic conditions and gases (Belias *et al.*, 2003) and bottom water oxygen can be depleted (Aubin, 2009; Folke and Kautsky, 1989). In rare cases, fish farming has induced reduced levels of dissolved oxygen in water which can contribute with other polluting sources to eutrophication leading to the formation of toxic algal blooms (Folke and Kautsky, 1989; Frankic and Hershner 2003).

As well as organic discharges, inorganic and chemical discharges can have negative impacts on water quality (Hambrey *et al.*, 2008: 218) and even on the food web, including from the use of veterinary medicines to treat disease, chemicals for parasites, disinfectants and antifouling paint used on the cages (Belias *et al.*, 2003; Folke and Kautsky, 1989; Mantzaurakos *et al.*, 2007; Read and Fernandes, 2003; Russell *et al.*, 2011). Scholars have also documented evidence of marine rubbish from support vessels (Frankic and Hershner, 2003).

Fish farming can also have negative impacts due to interactions with wild fish. A first such interaction is from fish disease outbreaks (Frankic and Hershner 2003), whereby fish farming can create conditions which are conducive to disease (Folke and Kautsky, 1989) and, in the case of salmon and seabream, for example, sea lice infestations (Adams *et al.*, 2012; Papapanagiotou and Trilles, 2001; Pert *et al.*, 2014). Both disease and sea lice infestations may spread to wild fish populations (Mantzaurakos *et al.*, 2007), where for example sea lice abundance has been linked to an increase in mortality of migratory wild salmon (Pert *et al.*, 2014).

Interactions with wild populations can also occur through fish escapees from the cages which can result in genetic interactions when escaped fish interbreed with wild stocks (Folke and Kautsky, 1989; Mantzaurakos *et al.*, 2007; Pert *et al.*, 2014). Fish escape when nets are damaged, for example, by seal attacks.[4] Scent from the cages can also potentially confuse wild fish migrations (Folke and Kautsky, 1989).

In the case of trout farming, because the trout farm requires the partial damning of the river to redirect water flow, if it is sited on a fish river run, it can create physical obstacles for migratory fish movements (both up and downstream) thus disrupting ecological continuity and preventing fish from reaching key spawning grounds. Finally, in the case of freshwater trout farming, there is an important impact on water quantity as well as water quality. Because trout farming diverts part of the river flow through the farm, it can affect overall water quantity in a river with potentially critical seasonal impacts, e.g. in the case of drought (Fourmond, 2000).

From this brief overview, it can be ascertained that fish farming can have both localised and wider ecosystem effects (Folke and Kautsky, 1989; Mantzaurakos *et al.*, 2007; Sutherland *et al.*, 2009). However, certain qualifications have been made in the literature concerning these negative impacts. First, a common refrain is that although aquaculture does have impacts, it 'often has a smaller impact than other human activities' (Costa-Pierce, 2010: 93; Soto *et al.*, 2008: 19). Second, aquaculture can also have positive environmental impacts – for example, through growing fish for restoration purposes and for the re-stocking of rivers following fish stock collapses (Soto *et al.*, 2008).[5] Third, it is acknowledged that its actual impact is dependent on how it is managed (Read *et al.*, 2001):

> Many of the negative impacts … [were] reduced or eliminated by appropriate management measures.
>
> (Pillay, 1997: 9)

This is because the

> environmental impact is dependent on fish density, feeding systems and geomorphology and hydrodynamic conditions in a fish farm area.
>
> (Belias *et al.*, 2003: 294)

For example, it has been argued that benthic community effects depend on hydrodynamics. Therefore governing rules which take either sea currents or river flow rates into account in the siting of farm are likely to lead to a reduction of these kinds of impacts (Folke and Kautsky, 1989). Well-flushed farm sites will minimise negative impacts (Ifremer, 2003). Hence, governing rules on siting are critical. Similarly, fish farming is believed to contribute only marginally to eutrophication, but this depends on the number of farms in any given area (Soto *et al.*, 2008), which again can be subject to rules governing these carrying capacity practices.

A common refrain, therefore, is that much depends on governing and management choices and that 'legal regimes' have a clear contribution to make to ensure that aquaculture development takes place within the 'carrying capacity' of the environment (Glenn and White, 2007). However, it is only through empirical examination of the institutional and political work of the actors who produced the legal frame that we can understand why a particular frame, and the institutions which support it, have come about. Consequently, in the rest of this chapter, I set out how actors in Scottish salmon, Aquitaine trout and Greek seabass and seabream have worked institutionally on fish farm/environment interactions establishing changing policy directions over time.

Governing Scottish salmon farm/environment interactions

Overview of the changing sustainability narratives

A central element to the story about the changing government of Scottish salmon farm/environment interactions has been the changing role performed by Environmental Impact Assessments (EIAs). The early years of the Scottish salmon industry were ones of growth and expansion through technical innovation, improved husbandry and production efficiency (Young *et al.*, 1999). Leases were granted to companies to grow salmon by the Crown Estate[6] and in discussion with the Scottish Office.[7]

Against this background, early institutionalisations of sustainability in Scottish salmon farm/environment interactions were dominated by the Crown Estate and the Scottish Office working with producers. These actors strongly tended to depoliticise ecology–economy interdependencies associated with farm/environment interactions. During this early period (1970s–2000s), although fish farming had environmental impacts, nonetheless it 'was perceived as a (relatively) environmentally benign activity to support community stability and regeneration' (Peel and Lloyd, 2008: 365). This can be understood in the context of a wider narrative which regarded the biophysical innovation associated with fish farming as a means of decoupling the provisioning of the seafood market from environmental degradation of wild fish stocks.

The first granting of leases by the Crown Estate to companies for the purposes of salmon farming only took into account minimal environmental considerations (e.g. siting of farms: Lloyd and Livingstone, 1991). Data on more comprehensive environmental impacts were only specifically requested following the entry into force in Scotland of EC Directive 85/337[8] necessitating that an EIA be carried out.

These early beginnings notwithstanding, by the end of the 1980s, this ecology–economy 'win–win philosophy' had come under attack. Environmental social movements such as the Scottish Countryside and Wildlife Link (today the Scottish Environment Link), Friends of the Earth and WWF (World Wide Fund for Nature) began to publish reports which both challenged sustainability premised upon this narrative and sought to disrupt it.[9] These attacks continued into the early 2000s, increasing pressure on the institution through clashes of sustainability narratives politicising a range of ecological impacts (Berry and Davison, 2001). In response, producer organisations fought back and political struggles which occurred around this time were conflictual:

> There was a lot of pressure to change the way government and industry and other stakeholders interacted because it was a very toxic relationship.
> (Interview: Scottish stakeholder organisation, 2011)

However, what I found was that, by the end of the 2000s, not only had this toxic relationship been defused, but the policy substance of sustainability had changed. Indeed, the early institutionalisation of sustainability had been overlayered with a new one endorsing anthropocentric ecosystem approaches problematising ecology–society interdependencies.

This transformation of the policy substance of sustainability was accompanied by important changes in the governing framework. First, following UK devolution, a newly elected Scottish Government (SG) and Parliament (SP) acquired policy-making powers in aquaculture which they used to develop a new Scottish strategy on the industry. Second, within the governing framework of a new sectoral policy strategy specifically dedicated to aquaculture,[10] a shift was made from the past 'privatised institutional context' (Lloyd and Livingstone, 1991: 263) towards a participatory governing process (Peel and Lloyd, 2008). The new governing framework consisted of Ministerial working groups opened up to producers, scientific organisations and other stakeholders meeting to discuss key problems identified in salmon farming, one of which was farm/environment interactions.[11]

In particular, those participating in these working groups now included some of the very same environmental-NGOs (e-NGOs) which had initially attacked the industry in the 1990s. Accordingly, whereas during the early creation of the institution farm/environment interactions were defined by dominant actors at the time as occurring at the scale of the farm site and depoliticised, the re-institutionalisation of sustainability by a new set of actors now re-defined farm/environment interactions at a wider ecosystem scale of the water body and problematised them. Over and above viewing salmon farming simply as a technological fix, actors now began debating aquaculture's relationship to nature in terms of a 'socio-ecosystem'. This anthropocentric ecosystem approach was, however, underpinned by important new tensions between utilitarian versus distributional theories of politics.

Unpacking interdependencies

Political work on the sustainability institution governing farm/environment interactions has been characterised by clashes at the frontier politics of inter-dependence and territory, political struggles between private and public regulation, and the depoliticisation of conflicts between knowledge types. Taken together, these have facilitated the destabilisation of one narrative of nature—society interdependencies by attempts to place another one at the heart of the sustainability institution governing the environmental impacts of fish farms and the interactions of production and biophysical processes.

Territory: The legitimacy of governing institutions is critical for their effectiveness. The contents of rules alone cannot explain their power. Rather, institutions have to be meaningful to those actors who are affected by them in order to exercise their authority. Throughout the 1980s and 1990s, salmon farm/environment interactions had been governed within a centralised UK-state, whereby the UK government in Scotland (the Scottish Office) and UK Crown Estate Commissioners authorised the leasing of sites to salmon farming. This organisation of territorial power became increasingly contested, however:

> [Salmon farming] was essentially governed from London until devolution. And then there was a fundamental change in the way that particularly aquaculture has been managed in that aquaculture was largely managed from London and there were officials in Scotland that had some say in the way that it would be conducted, but ultimately the decisions were made in London. Which was bad because these people had no experience of the industry; they weren't based in Scotland and so they were making decisions that were not necessarily appropriate. And that caused many problems.
>
> (Interview: Scottish scientist, 2011)

UK devolution in 1999 offered new material and imagined resources and identities to develop a distinct 'Scottish' policy. Policy competences over aquaculture were devolved to Scotland under Scottish devolution legislation – a competence which the incoming SG (and SP) would share thereafter with both the UK government and EU decision-making bodies. These new powers were quickly seized upon by a range of actors. Not only did the new SG[12] (and SP) view devolution as an opportunity to make a difference to the way in which aquaculture had been regulated, but a range of stakeholders wanted change in this direction, including producers:

> [...] the need for them to be more strategic about supporting our industry. Very early after the formation of the Scottish Executive[13] in 1999, we managed to get them to agree to form a working group – the Ministerial

Working Group in Aquaculture – I was one of the initial members, I still am.

(Interview: Salmon producers' association, 2012)

Pressure was brought to bear upon Scottish public actors to define a (new) Scottish territorial approach to regulate salmon farm/environment interactions. In 2003, an initial step was the launch of a dedicated Scottish sectoral strategy for aquaculture.[14] The first (Liberal–Labour coalition) SG in office published its strategy in 2003 and this was renewed in 2009 by the SNP SG. This strategy provided actors with a new policy space to discuss a range of issues related to the 'sustainable governance' of aquaculture in which salmon farm/environment interactions were central elements. This was a critical change, because up until then, aquaculture had been predominantly governed through transversal EU and UK regulation (for example, on pollution or use of medicines) and through 'temporary governors' (Georgakakis, 2012). The development of the new strategy would, for the very first time, establish permanent governors for salmon farming and also allow for the development of policy designed to address industry-specific issues (as distinct from general ones).

An important objective of the new approach was that its implementation would conform to core principles of devolution, namely that decision-making would be inclusive and transparent, and that public actors would engage with stakeholders. The implementation of the strategy was therefore structured around the setting up of a steering Ministerial working group, along with thematic working groups bringing together public and private actors (producers, feed companies, retailers, e-NGOs, wild salmon fisheries' associations) to discuss sectoral issues (we return to this below).

The operation of the working groups thereafter provided a new institutional space in which a Scottish 'interest' in fish farming could be constructed. This would be gradual and arise as actors' interests developed around a debated 'Scottish' vision of salmon farming. Of course, this is not to suggest that this was a shared vision. There were many conflicts in these groups. However, the working groups and the strategy did provide a governing space in which to problematise issues of salmon farm/environment interactions and bring to fruition a 'Scottish' politics regulating these problems.

Along with these efforts to create a separate Scottish policy, actors have also advanced institutional work to manage the boundaries of (newly defined) Scottish public action engaging in the UK and EU arenas. This has led them to pursue multi-positioning strategies to achieve coherence in policy practice. Actors have developed a number of parallel strategies of engagement, both seeking to influence UK and EU policies via domestic inter-governmental negotiation[15] as well as via their collective industry organisations.[16] In inter-governmental arenas, Scottish policy officials have particularly sought to demonstrate their authority over fish health issues, and have sought to be present in EU Council working groups on these very questions, at times taking the 'UK' lead in EU discussions.[17]

Other strategies for managing the new frontiers of Scottish public action have been to discredit EU approaches. For example, on interview, producers' social representations of EU governing approaches to aquaculture were particularly negative, compared to their view of Scottish ones, which they all praised. Indeed, one of the main arguments advanced by them during the 2007–2009 European Commission consultation on European aquaculture was that, from their perspective, no further EU public intervention was necessary to govern this industry:

> The most important arena is Scotland at the moment.
> (Interview: Salmon producers' representative, 2010)

Actors stressed the importance of Scottish 'stewardship' (Scottish Executive, 2003) in this regard. Indeed, in the social construction of post-devolution Scotland, actors frequently made connections between salmon farming and the Scottish natural environment within stories about Scottish territorial continuity. The following quotations capture the tone:

> [...] the salmon sector in Scotland which is worth over 400 million pounds a year to the Scottish economy, which is great and that is a growing figure, although it's not massive. But what is important about salmon farming and aquaculture more generally is where it is in Scotland – in the coasts and large number of rural communities.
> (Interview: Scottish public actor, 2010)

> What we do in farming is identical to what happens in nature.
> (Interview: Scottish salmon producer, 2010)

These kinds of images of Scottish relationships to nature were also expressed in attitudes towards animal welfare. For example:

> [...] high standard of animal welfare. Key to that is good staff, we trained them.
> (Interview: Scottish farm site manager, 2012)

> 60 per cent of Scottish salmon industry has the label "Freedom Food".[18] If all the Scottish industry had that, it would be fantastic as then Scotland would be looked at as the fish welfare-friendly nation in comparison to every other producing country in the world and that would give us a strong edge. And that is important.
> (Interview: Scottish salmon producer, 2010)

Consequently, in the building of new frontiers of territorial action, the ownership and responsibility for governing salmon farm/environment interactions has been taken out of the hands of a restricted partnership of Crown

Estate Commissioners and producers towards a wider range of Scottish public and private actors. This has been fundamental to the replacement of the usage of the EIA as a managerial technique aimed at salmon companies within an initial ecological modernisation narrative by its wider usage as a policy tool addressing ecology–society interdependencies. To bring this about, Scottish 'society' needed to emerge both as a category of thought and as a regulatory space.

Regulation: Cross-cutting and in interaction with public action on territorial interdependencies, actor powering around public/private regulatory interdependencies was also part of their political work re-institutionalising sustainability. Pre-devolution, rights for companies to set up a fish farm were allocated within a 'private management planning' process through leases granted by the Crown Estate which at that time comprised the seabed and foreshore around the UK coast (Lloyd and Livingstone 1991: 253) (see Chapter 5). Whereas, as described above, Crown Estate Commissioners had developed their own guidance on siting, EIAs first came onto the regulatory agenda in 1998 following the implementation of EC Directive 85/337 (Priyan and Smith, 1994). Over time, the contents of EIAs were gradually fleshed out to cover organic and inorganic impacts, whilst other UK general laws covered fish diseases.[19]

Post-devolution, the private nature of this planning system was called into question. For example, an e-NGO report published in 1988 suggested that few EIAs had been carried out up until this point (Scottish Wildlife and Countryside Link 1988), and over the next decade conflicts emerged over whether a full range of environmental impacts was actually known (Berry and Davison, 2001). Consultants argued that it was 'no longer sufficient to [just] abide by the [EIA] regulation' (Nautilus, 2002). According to these points of view, the EIA was a limited tool and did not allow for broader discussions on choices. Calling instead for a 'sector environmental assessment and management plan' (Nautilus, 2002: 8) (in line with a new EU directive),[20] they made the case for rendering the process more democratic, thus allowing for input from environmental pressure groups as well as local authorities.

In response to these multiple calls to open the process up to greater democratic scrutiny, the SP made a critical decision to grant local authorities the final say in planning authorisations. Whereas initial authorisations of sites were conducted within a governing philosophy whereby authorisations were viewed as providing leverage for growth under minimal application of EIAs, the shift towards democratising the planning authorisations also brought about another shift towards a strengthening of EIAs, including a change of scale in implementing ecosystem approaches. This was because as well as re-assigning authority from private to public actors, the SG also strengthened and tightened up the role of its key regulatory agencies (namely, the Scottish Environmental Protection Agency [SEPA] and Scottish Natural Heritage [SNH]). These regulatory agencies would provide statutory advice on applications' farm/environment interactions.

SEPA was formed bringing into it different people from the local authorities and the river boards and then it was decided that there was a need to
have a very co-ordinated way of regulating aquaculture from the pollution point of view. That set up the development of aquaculture specialists
and the development of the fish farm manual, which drew together all
the policies and best practice into one easier to follow manual.

(Interview: Scottish public actor, 2010)

SEPA's role has been to provide (or not) a discharge consent licence regulating discharges from farms.[21] Allowable environmental impacts are monitored
every two years concerning both seabed impacts and chemical residues on a
site by site basis and 'allowable zone of effects'. The implementation of EU
Directive Natura 2000 also further required that EIAs be used to monitor
farm impacts on biodiversity and protection of named flora and fauna. Advice
on these impacts has been provided by SNH. Finally, in 2007, the SG government took this process further and created a new division called Marine
Scotland which monitors the water body as a whole and cumulative effects of
different farm sites, as well as the impact of fish diseases and fish health issues.
Taken collectively, these regulatory bodies therefore govern an extensive
range of organic and non-organic impacts and interactions with wild fish
populations at the ecosystem scale.

This strengthening of public control over planning permissions, and con-
comitant extension of the reach of public authority over EIAs, was also
accompanied by an extended role for private stakeholders in decision-making.
Whereas in a first move the private approach to planning had been problematised and replaced with a public one, in a second move complementarities
between public and private regulation were sought.[22] This happened through
the setting up of working groups in 2009 bringing together stakeholders to
discuss controversial issues in fish farm/environment interactions.

Two of these working groups can be briefly mentioned here: the working
group on fish health and the working group on containment. Whereas by the
late 2000s organic and non-organic impacts were mostly regarded as 'governed', farming's interactions with wild fish remained controversial. An issue
seen as critical was the problem of sea lice, which has not only been a
problem for the industry, but has also been a problem for wild fish populations. Sea lice outbreaks had led to the setting up of voluntary Farm Management Agreements (FMAs) between different companies sharing the same
water body. Under FMAs producers had agreed to only farm fish of the same
generation at any one time in the water body and to have fallowing periods
to diminish the spread of disease. However, FMAs had been established as
voluntary agreements, thereby weakening their effectiveness because not all
producers were members. Conflictual discussions in the fish health working
group (especially between producers and representatives of wild salmon fisheries' boards) led to proposals that FMAs become obligatory – and this was
enacted in the new Scotland Fisheries and Aquaculture Act in 2013.[23]

Members of salmon fisheries boards continued to be critical of producer practices, however:

> [T]he impacts on wild fish demand a standard of lice management that is more rigorous than they would perhaps set for themselves. So that has not been working so well.
>
> (Interview: Salmon fisheries boards' representative, 2012)

Nevertheless, they have welcomed these moves, along with other initiatives towards sea lice control,[24] which have only been made possible through these public/private negotiations. A second controversial issue has been 'containment' of the salmon in the net cages, with many salmon escapes documented every year. The second working group has also led to action finding complementarities between private and public regulation. Producers and bureaucratic actors both considered that a regulatory response was required:

> The containment working group, we had a huge breakthrough there in that the industry has come back to government and said we think that we should have a statutory engineering standard for fish farms.
>
> (Interview: Scottish public actor, 2010)

This was also brought within the Scottish Fisheries and Aquaculture Act, whereby a new engineering standard has been implemented which 'forces' technological innovation in net design to seek to reduce numbers of escaping fish.

Knowledge: Finally, important political work on knowledge conflicts has contributed to discussions on sustainability's social meaning. Initially, to destabilise the win–win ecology–economy sustainability institution governing farm/environment interactions, e-NGOs had sought to politicise these interactions and highlight knowledge gaps claiming that these impacts were not fully known. However, in the re-institutionalisation of sustainability problematising ecology–society interdependencies, a reverse strategy was put in place by the 'new' governors. More specifically, science and knowledge has been mobilised to depoliticise controversies.

First, regulatory agencies invested heavily in new regulatory science. This has led, for example, to the development of a new ecosystem model used by SEPA to judge and monitor discharges – the DEPOMOD model (Cromey *et al.*, 2002). Similarly, Marine Scotland has worked to develop and update its models on water quality and fish health, modelling at the scale of the water body:

> Aquaculture development must be within the assimilative carrying capacity of the water body in which it is situated. It is important to demonstrate that water bodies are able to absorb the waste products associated

with aquaculture without harming the marine or freshwater ecosystem. This process underpins Marine Scotland's Locational Guidelines for Aquaculture as well as advances in water body scale modelling that are currently under development. Such an approach is also consistent with the Water Framework Directive and the ecosystem approach required by the Marine Strategy Framework Directive.

(Scottish Government, 2009: 5)

Investment in regulatory science has led its proponents to advance narratives showing that whereas in the past the environmental impacts of salmon farming suffered knowledge gaps, today the impact is knowable, known and is being monitored.[25]

Second, investment in regulatory science has been accompanied by collective investment in new areas of research to support policy. The first Ministerial working group established in 2003 set up a Scottish Aquaculture Research Forum (SARF) whose commissioning board brings together all the different public and private actors of the industry. These actors collectively decide which areas of research they wish to fund and who will carry out the research. This has led to the development of collective ownership of different types of research, leading in some cases to the development of regulatory instruments, e.g. the engineering standard mentioned above.[26] SARF has funded many projects looking for alternative solutions to managing fish farm/ environment interactions, e.g. on non-chemical solutions to sea lice through the use of feeder fish.

Third, the SG has striven to render data on aquacultural environmental impacts transparent. For example, both itself and the regulatory agencies

Table 4.1 Summary of political work over Scottish salmon farm/environment interactions

Scottish salmon			
Social constructions of sustainability	From de-politicised ecology–economy interdependencies to anthropocentric ecosystem approaches problematising dynamic ecology–society interdependencies		
Frontier politics of polity interdependencies	Territory	Regulation	Knowledge
	Separation of Scottish approach and Scottish societal ownership; New identities and jurisdiction; Multi-positioning strategies on UK, EU scales	From private guidance on EIAs to public control; New public sectoral policy complemented by new private regulation	Claims of knowledge gaps to de-stabilise initial institutionalisation; Scientification of controversies in re-institutionalisation; Conflicts over interactions between farmed and wild salmon

publish data on farms on a new website which has been set up for this very purpose.

However, strategies to depoliticise fish farm/environmental interactions through the scientification of controversies has not gone unchallenged. In particular, e-NGOs not engaged in the working groups have continued strong media campaigns to publicise what they perceive as 'secret data'. This has been acquired through requests for access to information (e.g. emails) under the UK Freedom of Information Act, which have then been published on their website. fairly One of their more recent attacks was the publication in January 2017 of the increasing levels of chemical treatments by salmon companies, which not fit well with a proclaimed public policy of reducing the environmental impacts of chemicals.[27]

Governing Aquitaine trout farm/environment interactions

Overview of the changing sustainability narratives

In the case of the sustainable government of Aquitaine trout farm/environment interactions, here too a central element in the story has been the role played by EIAs in relation to the EU Water Framework Directive (WFD). Unlike in the case of Scottish salmon, however, where ecosystem approaches were mobilised in the 2000s to replace depoliticised ecology–economy interdependencies, the first period of growth of Aquitaine trout took place within a general French-wide legal regime which had already embraced an ecosystem approach. This was because trout farm sites in Aquitaine required authorisations both for their installation[28] and their river water use,[29] thus bringing their regulation within a broader framework of French water governance. This framework had already been applying an ecosystem approach to water governance within territorial anthropocentric applications since 1964 (Bouleau, 2014; Bouleau *et al.*, 2016). Both water quality and water resource use were defined at the scale of the watershed, whereby authorisations for water use applied norms of equity which took into account all stakeholders and economic activities using the same stretch of water (Fourmand, 2000):

> You have to leave water in the river for other usages and for the biological life of the river. ... This is even a core principle of civil law, you have to leave enough water for downstream users, you know! Otherwise, you are doing a 'Jean de Florette'!!
>
> (Interview: Ministry of Ecology official, 2012)[30]

The first institutionalisation of sustainability of trout farm/environment interactions was thus predominantly shaped through administrative processes dominated by public actors. In particular, departmental prefectures, in consultation with other regulatory bodies acting on the watershed,[31] authorised farm installations on the basis of EIAs.[32] These were further amended by the EC Directive 85/337 and operated from within the general principle that whilst trout farming required good quality water to grow the fish, it must also restore water quality to the river (Fourmand, 2000).

> Water belongs to everyone. [There is a] responsibility for all users to col-
> lectively respect the good ecological status of water bodies.
> (Interview: French public actor, 2012)[33]

By the mid-2000s, however, anthropocentric applications of the ecosystem approach were being challenged by a more ecocentric one affecting trout farming. On the one hand, local environmental movements, such as ANPER-TOS,[34] began to attack prefectural authorisations in administrative tribunes on the grounds of perceived negative environmental impacts of trout farms. On the other hand, the implementation of the EU WFD in France[35] re-institutionalised state-wide images of rivers. Whereas in the past river water quality had been conceptualised as coherent with economic activities and 'living rivers' (Germaine and Barraud, 2013), now some public actors[36] implementing the WFD[37] expressed uncertainty over whether this would still continue to be the case, evoking images of 'wild' rivers over 'living' ones (Germaine and Barraud, 2013):

> It's a massive problem for biological life if the river is not free flowing
> with places to meander, with rapids, deposits, zones of free circulation
> and sedimentation, all of which means that this is a living environment.
> [Instead] it is straightjacketed by a succession of small weirs which rather
> creates a succession of ponds, rather than a living flowing river.
> (Interview: French public actor, 2012)[38]

Arguments over rivers created tensions over the social meaning of sustainability governing farm/environment interactions and uncertainties for trout farmers. This was especially because the definition of the contents of this institution (i.e. water resources) had mainly been decided through actor political work over *general* water regulation at both EU and state-wide scales, in which neither trout farmers nor officials from the fish farming division of the French government[39] had been involved. Further, whereas these general rules applied to managing inter-sectoral conflicts along a watershed, the absence of a specific French sectoral policy or dedicated governance framework[40] for fish farming seriously weakened fish farmers' influence over the contents of the general rules. Instead, producers' associations found themselves at the margins of a wider political community governing water resources:

> When implementing the WFD we reasoned globally and not by indi-
> vidual activity.
> (Interview: French public actor, 2012)[41]

To try to influence the content of sustainability as an institution governing farm/environment interactions towards more anthropocentric applications respecting WFD norms, fish farmers began to conduct political work through their collective private sectoral associations working in public/private partnerships, especially within the Aquitaine region, to accord their production what they referred to as a 'clean environmental identity'. Through these (and other) initiatives, attempts were also made to raise awareness on trout farming's sectoral issues when implementing the WFD. This political work was not aimed at disrupting the sustainability institution governing trout farm/environment interactions, but at (re-)institutionalising it within anthropocentric applications.

Unpacking interdependencies

Public action over Aquitaine trout has included political work to legitimise industry actors' territorial visions of trout farming's interactions with the environment versus state-wide general visions upheld by actors governing water, sector-specific public/private co-regulation as a counterpoint to general public regulation, and the generation and mobilisation of new types of knowledge.

Territory: Since the early creation of trout farms in Aquitaine in the 1980s, it has been the French state which has held political authority governing trout farm/environment interactions. Authorisation for licences are granted by the prefecture. Within this procedure an administrative judge is appointed to review the EIA submitted, consult with the relevant public regulatory agencies,[42] and organise the public inquiry through which different stakeholders may engage. At the end of the process, the prefecture ultimately gives its consent (or not) to the application.

The territorial organisation of political authority underpinning this procedure has not fundamentally changed since the 1980s, whereby both the contents of EIAs continue to be set through general French-wide state regulation and it is the 'state' acting through the prefecture which authorises the licence. Indeed, although the Aquitaine region holds potential policy-making powers over aquaculture, successive regional councils have not engaged in conflicts with the central state to assert political authority over this sector. As argued by Pasquier, (2015), French regionalisation carried with it the potential to disrupt the previous French public policy style as described by Jobert and Muller (1987), whereby one dominant actor set the referential of the public policy. This potential has been realised over some public policies (e.g. on integrated maritime policy [Saliou, 2010]), however, there has been no evidence of this form of regionalisation of public policy in respect of trout farming in Aquitaine. Instead, the focus of the region has been rather on funding mechanisms and providing financial support to the industry in a logic of territorialisation of socio-economic resources rather than political power (Douillet, 2003).

Against this background of non-politicisation of the boundaries of public territorial jurisdiction over trout farming, collective private actors have, however, mobilised around Aquitaine territory as a category of thought to seek to influence a (public/private) regional approach. In particular, working through the collective organisation of GDSAA,[43] they have lobbied the regional council and joined forces with it to develop important public/private initiatives setting out a 'regional' collective view on the contents of trout farm/environment interactions. GDSAA was initially set up in 1984 to work on pollution, disease and fish re-stocking practices, forging an alliance between recreational fishers and trout farmers around these issues. GDSAA has worked at the joint territorial scales of both Aquitaine and the watershed, developing good practice for trout farmers managing fish health issues. This has led to the development of fish disease-free 'identity cards' awarded to trout farmers. Critically, through GDSAA's political work, large areas of the regional river basins have been declared disease-free zones under the rules of the EU Fish Health Directive.[44]

Following controversies over the implementation of the WFD and working with the regional council, the 'Coop de France Aquitaine' and AFNOR (a commercial certification body), GDSAA developed in 2010 a new public-private charter, this time on environmental issues, called AQUAREA (l'Aquaculture Respectueuse de l'Environnement en Aquitaine).[45] This qualification programme allows trout farmers in Aquitaine to be certified for high environmental practices which 'prove' their sustainability according to a range of criteria governing all forms of ecosystem environmental impacts. The aim is that over time a vast majority of trout farmers in Aquitaine will be certified, thus contributing to a regional statement on the compatibility of fish farming with ecosystem functioning:

> Working through AQUAREA ... this enabled producers to have individual information on their own farming practices and at the same time to have a collective means of communicating with the administration to defend their interests.
>
> (Interview: Trout producers' representative, 2012)[46]

In this territorial work, as was the case in Scotland, actors have sought to naturalise trout farming's relationship with the natural environment in line with territorial continuity. On interview, actors evoked multiple notions of territory; as well as Aquitaine, they evoked 'les Landes', 'the Pyrenees' and the 'Basque country'. At times this positive relationship was expressed in comparison with other regions:

> In the Basque country, we have always been 'clean'.
>
> (Interview: French trout producer, 2012)[47]

In other regions, environmentalists have built negative perceptions of fish farming, but perhaps fish farmers have also worked poorly.

> (Interview: French trout producer, 2012)[48]

Whereas Scottish actors discussed notions of 'stewardship' linking ecology–society interdependencies, French producers have also sought to stress their societal role to legitimise their arguments by constructing alternative interpretations of the ecosystem approach. Representing their interactions with the biophysical in terms of their societal role as 'les sentinelles' (along the lines of being a canary in a coal mine), they have sought to naturalise their economic activity as fitting in with nature. The concept of the 'sentinelle' was both institutionalised in AQUAREA and evoked on interview. In the past, trout were used as indicators of water quality, as any change in the quality of the water could immediately be seen in the trout. Farmers have therefore argued that the trout could be used to monitor the quality of the water on an hourly basis:

> Before this the trout acted as a sentinel animal.... The food has to be clean at the place you farm fish. If there are people putting crap in the river ... my fish will die.
> (Interview: French trout producer, 2012)[49]

> Being environmentally friendly ... well we all do it ... But I also need good quality water otherwise my fish will not grow.
> (Interview: French trout producer, 2012)[50]

Producers also stressed their constant presence on the trout farm:

> In mountain areas in the Pays Basque, every fish farm has someone working all night ... We have people working 24/24 on the site. It is annoying, but if you are not there and something happens you can lose the whole lot.
> (Interview: French trout producer, 2012)

And their very strong attachment to the environment:

> We are hooked on the environment, really hooked like on drugs, and we depend on upstream water and we depend on downstream water because legally we have to ensure minimum discharges in order to have our licence to continue using the water, so for us the environment, well we live in it ... we bathe in it.
> (Interview: French trout producer, 2012)[51]

These very strong attachments to local nature and territory have led to political mobilisation at the regional and French-wide scales to engage with the setting of French-wide rules. This political work has been carried out through the GDSAA, the South West Syndicate and through the inter-professional organisation, CIPA.[52] Yet, whereas producers have worked at the frontiers of territory within France seeking to converge their interests with those of others, they have experienced challenges engaging at an EU scale:

[The] problem comes upstream ... We are completely absent at the European level – Clearly.

(Interview: French trout producer, 2012)[53]

This has contributed to their vulnerability in defining sustainability in the case of the implementation of the WFD in France, and is particularly marked when EU policy is presented by the French government as being a 'fait accompli' and non-negotiable. For example:

We have European texts and the rules are very clear concerning our obligations to restore migratory fish. These has been imposed by Europe.

(Interview: French public actor, 2012)[54]

For all these reasons, overall responsibility for governing trout farm/environment interactions has remained by default with French state-wide (extra-sectoral) actors governing water through transversal regulation (EU and French).[55] This has created tensions within territorial interdependences which have been worked on by producers seeking to give themselves the opportunity to take responsibility over their very destiny and to shape the sustainability institution in line with their own narratives. However, they have also experienced these tensions as a constraint, instilling feelings of being victims of the decisions made by others in distant places.

Regulation: Political work at the interface of public and private regulation reveals similar ambiguities. As we saw above, since the 1980s, trout farm/environment interactions have been governed through state-wide public rules on water quality and use. These have been strengthened over time so that EIAs now govern a full range of organic and inorganic impacts as well as interactions with wild fisheries (e.g. through implementing EU policies such as Natura 2000, WFD, Fish Health Directive, etc.).

More recently, the implementation of the WFD has caused controversies over its consequences for trout. These controversies have centred in particular upon uncertainties over the way in which the Good Ecological Status (GES) of water will be monitored. Points of concern have included the measurement of discharges from farm sites and how to manage fluctuations in water levels in the rivers with acceptable levels of discharges;[56] the cost of the modification of trout farm weirs to allow for ecological continuity; how to meet low-water flow targets in high season (Fernandez and Debril, 2016) whilst at the same time having enough water for the fish;[57] and fears that ecological continuity will allow for fish diseases to recirculate throughout the river, thus negating extensive efforts made within GDSAA to render large areas of the watershed disease-free zones. Animating these controversies have been a complex set of arguments put forward both by the Ministry of Ecology and by producers. These turn especially on whether or not the efforts being demanded by the French government of trout farmers in meeting the GES are disproportionate to those being asked of other river users.[58]

Whilst debates over implementation issues are not unexpected, the main challenge which industry actors have faced is the absence of a sectoral governing arena in which issues of this kind can be collectively discussed. Instead, in the case of trout farming, public/private negotiations have been conducted in classical French ways by lobbying the Ministry or by seeking private meetings between collective inter-professional associations (e.g. CIPA) and Ministry officials. Moreover, the character of these public/private exchanges has been at odds with broader changes made in France since 2007 on governing approaches regulating the environment. Indeed, an important shift in the mode of governance had been made following extensive stakeholder consultation over the implementation of the state-wide environmental programme 'Grenelle' (Lascoumes, 2011). This has further entailed efforts to replace the former 'regalien' or 'top-down' style of governance with new mechanisms institutionalising a participatory eco-democracy (Boy *et al.*, 2012).

Of course, this participatory approach has been applied to water governance more generally – and trout farmers can and do participate in river basin committees. However, in these arenas, they are in a small minority. Consequently, for these reasons too, trout farmers have engaged in public/private charters and other private self-regulatory instruments (e.g. IDAqua; see below) to self-regulate their farm/environment interactions. In this manner, producers working with regional public actors and research institutes have sought to create a separate regulatory space for trout farming in which they can generate their own meaning of farm/environment interactions. Private self-regulation is therefore capturing policy in a governing sense, creating a public/private sectoral exchange which is otherwise absent in the public sphere.[59]

Knowledge: Finally, conflicts within interdependencies of territory and public/private regulation have also taken place between different knowledge types. Legitimate knowledge governing the contents of the sustainability institution has been regulatory ecological science, which has dominated both the contents of the EIAs and the setting of indicators judging GES under the WFD (Bouleau and Pont, 2015). Expert scientific knowledge has dominated since the beginning and there has been no real destabilisation of this hierarchy. Yet this situation does not imply full consensus. On the one hand, there has been a problem of communication between actors and how they construct knowledge. This has been illustrated in state practices relating to monitoring water quality in farms. Whereas for the administration monitoring has been constructed as a question of having the necessary expertise to carry out the work correctly (and hence obtain an accurate measurement), from the point of view of the trout farmer this has been constructed as a social moment in which different knowledge and views can potentially be exchanged:

Experts from ONEMA[60] working on a river site have the necessary expertise to make a report on its normal ecosystem.

(Interview: French public actor, 2012)[61]

ONEMA does the control and they don't come by very often – they come when they want and we don't always see them. That has changed. When they came in the past, they stopped and said hello. Now they are overworked. They come right past my window and take the water and go off and they don't even say hello.

(Interview: French trout producer, 2012)

These differences in approach to knowledge acquisition have led trout producers to conclude:

They live in another world, they are not like us.

(Interview: French trout producer, 2012)[62]

On the other hand, trout producers have worked to develop their own knowledge as a counterpoint to official expertise.[63] They have, for example, set up partnerships with research institutes, especially with researchers from the Agronomic Research National Institute (INRA), ITAVI, the National Research Institute of Science and Technology for Environment and Agriculture (IRSTEA) or the French Research Institute for Sea Exploitation (Ifremer), to produce data on the industry.

This has led to the development of sustainability indicators through the creation of the tool IDAqua[64] (Gueneuc *et al.*, 2010) and the conducting of studies with feed companies to change the calculation of organic impacts made by authorities, accusing them of using out-dated feed conversion rates from 20–30 years ago. Through the application of their tools they have generated new data:

But at the same time, thanks to our collective association (GDSAA) the administration gets access to information not available in any other region.

(Interview: Trout producers' representative, 2012)[65]

Producers have also mobilised around their 'savoir faire', which they consider is under-utilised in discussions about farm/environment interactions:

And then on a daily basis, this demands important skills to manage all of the water's parameters ... It demands ready availability, passion and a great skill. They are not trained engineers but they are permanently gathering data in order to be able to make progress.

(Interview: Trout producers' representative, 2012)[66]

Table 4.2 Summary of political work over Aquitaine trout farm/environment interactions

Aquitaine trout			
Social constructions of sustainability	Initial anthropocentric ecosystem approaches are increasingly challenged by quasi-ecocentric ecosystem approaches governing water quality		
Frontier politics of polity interdependencies	Territory French state-wide approach (to water policy) dominates; Conflicts between public actors' state-wide visions versus trout producers' local territorial visions of rivers	Regulation Sector-specific private self-regulation as counterpoint to dominant public rules	Knowledge Dominant regulative ecological science; Generation of self-monitoring data as potential leverage for new representations of impacts; Conflicts over interactions between farmed and wild trout

> They [officials] want to treat the problem with the river, but they have only been to see it two or three times.
>
> (Interview: French trout producer, 2012)

As was the case for the politics over the interdependence of territory and regulation, tensions within the politics of knowledge interdependence have caused ambiguities over producers' positions acting in regard to the sustainability institution. To date, they have not changed their assignment of authority over knowledge, and the use of knowledge in institutional positioning strategies has not been consistent; often their expertise has been ignored or simply not invited (Tanguy Report, 2008).

Governing Greek seabass and seabream farm/ environment interactions

Overview of the changing sustainability narratives

The central element in the story about changing Greek governance of seabass and seabream farm/environment interactions has been around the very question of the licence to grow fish. Controversies over licences have in turn spilled over into controversies about EIAs.

From the start, interactions between Greek seabass and seabream farming and the environment were institutionalised through depoliticisations of ecology–economy interdependencies assuming that aquaculture could grow with minimal ecological impacts. However, this way of regulating sustainability became increasingly destabilised through attacks made by local communities. These actors have increasingly sought to disrupt claims to sustainability as put forward within this narrative and to open up otherwise depoliticised ecology–economy interdependencies to political contestation. In response, Greek governmental actors have sought to elaborate new siting regulations for licences premised upon ecosystem concepts. These new policy tools notwithstanding, however, the ecology–economy win–win philosophy has remained dominant in the regulation of farm/environment interactions.

In more detail, the institutionalisation and re-institutionalisation in Greece of farm/environment interactions depoliticising ecology–economy interdependencies has taken place in a situation of 'legislative anarchy' (Conan, 2000: 274). This situation has arisen through the non-application of legal texts, non-compliance with the law and myriad contradictions between actor representations of the 'problem', which all feature as characteristics of sustainability public action.

The governing of seabream and seabass farm/environment interactions has resulted from a layering of laws, decrees and Ministerial guidance enacted over the years. Environmental protection has been a constitutional principle in Greece since 1975 (Conan, 2000).[67] This principle, along with those contained in EC Directive 85/337, was translated into law in 1986 rendering EIAs obligatory for farm authorisations (Fousekis and Lekakis, 1997).[68] Further guidance on authorisation procedures was put in place in 1994 following Ministerial direction (Conan and Prieur, 2000) and these, and other laws, have been updated to apply a range of EU directives addressing environmental and fish health impacts and seeking to improve the site selection of farms.

The allocation of both production quotas and licences according to these environmental norms has been the shared responsibility of a number of public actors involving different ministries within the central government, their regional offices, and prefectures.[69] However, for many years some of these laws were not fully applied due to an absence of enabling legislation. For example, Conan found that, ten years after its publication, the necessary legal application measures for the 1986 environmental law had not been forthcoming and that authorities had to rely on other instruments to protect the environment (Conan, 2000). Additionally, also during the 1990s, non-compliance with environmental rules on carrying capacity by producers was commonplace.

Against this background, local communities attacked the licences in local tribunals and this eventually resulted in the Council of the State annulling licences (Karka, 2011). This being said, these permits were not overturned because of negative farm/environment interactions but for other reasons connected with their legality in the context of spatial planning (these points are discussed in detail in Chapter 5):

> Nobody lost their licence in Greece because they polluted the area. Maybe they lost it because someone wanted it to be a touristic area. But nobody lost it for polluting the sea. Not even one case.
>
> (Interview: Greek seabass/seabream producer, 2012)

The above notwithstanding, during this period the mobilisation of local communities, at times in alliance with e-NGOs, has politicised what communities perceive as farming-related pollution of sea lagoons. Yet, according to both government officials and producers (interviews), EIAs were conducted during this period and measurements taken, with the EIA being reformed in 2009 to render producers' carrying capacity more transparent. Indeed, international reviews of Greece carried out within the FAO (Food and Agriculture Organisation of the United Nations) found that 'the area under cage cultivation of seabass/bream was stable' (Costa-Pierce, 2008: 94) and that there was 'little evidence of adverse impacts, especially pollution from cages on a regional scale'. Scientific publications have also confirmed acceptable farm/environment interactions (e.g. Mantzaurakos *et al.*, 2007; Papoutsoglou, 2000).

Consequently, even though both public and collective private actors' depoliticisation of ecology–economy interdependencies has been destabilised through controversies over non-compliance with carrying capacity rules and negative environmental impacts, ultimately it has neither been completely disrupted nor replaced by an alternative construction of nature–society interdependencies as mooted by an alternate alliance of actors. What I found instead was that ecosystem approaches upheld within EU directives (e.g. Natura 2000), together with new local rules on carrying capacity, were being brought within the institution through their being applied within the EIAs. However, this process was not being carried by a clear alliance of actors seeking to transform it.

Unpacking interdependencies

Political and institutional work maintaining the ecology–economy win–win philosophy at the heart of the sustainability institution has taken place in relation to a lack of mobilisation around the frontier politics of interdependence and territory; a general absence of institutional work on private regulation (except for small companies accessing niche markets); and the emergence of new (and contested) knowledge types. This has also been the case for political work seeking to destabilise this social meaning of sustainability and replace it with an ecosystem one.

Territory: Whereas seabass and seabream farm/environment interactions are governed by a complex number of public bodies through multiple regulatory instruments set at different territorial scales, I found no evidence of political work seeking to open up these territorial interdependencies to controversy. Rather, any public action observed was more directly aimed at seeking to find synergies.

First, and as was also the case for Scottish salmon and Aquitaine trout, EU public general regulation covering EIAs, biodiversity, fish health, etc., governs seabass and seabream farm/environment interactions. However, by contrast with Scottish salmon and Aquitaine trout, both public and private actors interviewed were very positive about the influence of the EU on environmental and health policy which, they claimed, had provided new resources to build the ecologisation of the economy and the sustainability institution:

> Since Greece is part of the EU – and I hope it stays.
>
> (Interview: Greek seabass/seabream producer, 2012)

This conceptualisation of the EU was consistent with Greek sustainability policy more generally. As documented by others, Greek sustainability policy has been focused on the integration of ecological concerns with economic ones to permit the uptake of EU funding, which has been pivotal for Greek growth (Fousekis and Lekakis, 1997). This situation has also been the case for seabass and seabream farming which has benefitted considerably from EU funding.

Second, since the beginning, the dominant governor of seabass and seabream farm/environment interactions has been the Greek state, whether through its central or de-centred actors from the regional prefecture. Although the role of the central state governing environmental impacts has been criticised – for its complexity, its failure to monitor impacts or for its overall weakness in controlling farms – there was no evidence that these criticisms had been translated into political work aimed at territorial re-assignments of political authority over farm/environment interactions and policy processes. On the contrary, the overall tone of actors' comments went more in the direction of seeking to show the congruence between seabass/seabream farm/environment interactions and Greek approaches to the environment. These were depicted as being respectful of both the sea and animal welfare:

> We are not a factory that I don't care if I throw my waste into the sea, so you have to put me very high standards. Everything we have is in the sea. It is an open sea. I want to take care, more than you; I want to measure it; if someone one kilometre away is putting waste in the sea, or medicines.... I want to take care of it. Let's take care to protect my production.
>
> (Interview: Greek seabass/seabream producer, 2012)

> We respect environmental quality. Our products are the best in Europe; they respect strict controls.
>
> (Interview: Greek public actor, 2012)

The alternative territorial category which did emerge in the interviews, especially with producers, was that of the Mediterranean. Producers considered they were 'in connection with this sea' and expressed concern of the joint vulnerability of the sea and their business:

> Maybe a problem comes to Spain and then it goes to Greece … The Med is like a lake and in two or three years it will travel to us. We have to see it differently and we are building our association and also with other countries with the EU to see how we can face such problems in the future.
>
> (Interview: Greek seabass/seabream producer, 2012)

> There is an uncertainty every day. But this is also true of salmon – it is in the open sea. No matter how much you invest it is out there. It is out there.
>
> (Interview: Greek seabass/seabream producer, 2012)

> Because everything changes. The weather is changing and you have to be aware of new diseases which may come – all this brings some uncertainty.
>
> (Interview: Greek seabass/seabream producer, 2012)

For all these reasons, even though the authority of the Greek state has been called into question, political work has instead gone in the direction of the renewal of Greek state authority by addressing its weaknesses, rather than a re-assignment of political authority to another administrative scale.

Regulation: This strong tendency to depoliticise any tensions across territorial interdependencies has carried through into political work at the interface of public and private regulation. As I described above, actor struggles over the appropriate reach of public authority in relation to civil society have taken place within a context of 'legislative anarchy'. Seabass and seabream farm interactions have been governed mainly by public regulation. Initially, these licences were issued as government leverage to grow the industry, whereby the government had a starting quota of 320 licences to be distributed to companies wishing to set up in seabass/seabream farming. Over time, the EIAs demanded by the authorities have been strengthened. This has not only come about through their renewal following the implementation of EU directives, but also in response to controversies over non-compliance with rules on the carrying capacity of cages. Today, EIAs in Greece, as in Scotland and Aquitaine, cover organic and inorganic impacts, fish diseases, biodiversity, water quality, water chemistry, sedimentation of the bottom, flora and fauna of the area:

> The restrictions come from national and European legislation. All these restrictions form the legislative foundation to make the firm. From this legislation, we have the water quality measurements the farm has to do, they have to submit their measurements, keep their books, submit them to the competent authority so they must have the record also for the environment, it is the same for the EIA.
>
> (Interview: Greek scientist, 2012)

Consequently, public rules have dominated the regulation of seabass and sea-bream farm interactions. Indeed, although most producers have signed up to integrated management schemes (e.g. the International Organisation for Standardisation - ISO) as part of in-house company policy, there has been very little voluntary collective private self-regulation beyond an industry code of best practice drawn up within the collective producers' association – the Greek Federation of Maricultures (GFM).

This finding appears coherent with general public/private interdependencies in Greece. As research on other domains has shown, these had been practically fused together for many years in a state characterised by strong central interventionism coupled with a strong societal individualism (Papadakis, 2012). Only recently has a shift towards a political autonomy of civil society been observed (and I document this shift further in Chapter 5). Rather, a common strategy adopted to manage public/private interdependencies has simply been non-compliance with the rules. Non-compliance has occurred concerning the carrying capacity of the farm sites. Farms have produced above the tonnage limit as set by the licence[70] and have grown seabass and seabream in cages whose licences were for the growth of new species. Interviewees were very open about this:

> The problem is that 10 years before there was anarchy in the development of aquaculture and everybody did whatever he wanted.
> (Interview: Greek seabass/seabream producer, 2012)

> Because they recognise that they are already producing more than is on their licence. If they have the possibility to produce 250 tonnes, they are producing 500 tonnes.
> (Interview: Greek public actor, 2012)

> The approved licence for production in Greece must be between 70 and 80,000 tonnes. The real production in Greece is 120,000 tonnes. So you can understand how this expansion was realised ... over-production of some sites.
> Q. And everybody does it?
> Everybody.
> (Interview: Greek seabass/seabream producer, 2012)

This non-compliance has led to a loss of trust by local communities both of producers and of the state and suspicion over seabass and seabream farm/environment interactions. In turn, this has fuelled perceptions of fish farming as being 'out of control' and of polluting farms (because not controlled). It is not just producers who were not trusted. The state too lost trust as a credible governor depoliticising ecology–economy interdependencies. This was because it was argued that the Ministry of the Environment had not carried out all the checks it should have done in the past.

However, towards the end of the 2000s this situation began to change. Due to the need to regulate the planning of access to fish farm sites (which is described in Chapter 5), central ministries began to be more rigorous through issuing fines, forcing sales of fish and ultimately closing sites. Additionally, and in keeping with the beginnings of a possible transformation of the Greek state towards its political autonomy from civil society (Papadakis, 2012), central ministries launched a public consultation process with producers, e-NGOs, scientists and the public. This eventually led to the development of a new formula on carrying capacity to reduce non-compliance. This was because the new formula would be capable of capturing variations between sites in relation to the depth of the sea, currents and distance from the shore:

> I am operating a farm with a depth of water of 100 metres and he operates with a depth of 20 metres – why should we produce the same? And some other studies of universities came up saying that the carrying capacity of a farm should be according to the openness of the gulf, the depth of the water, currents and all of this. The new legislation gives you the ability to adapt your production according to the carrying capacity – so it's better.
>
> (Interview: Greek seabass/seabream producer, 2012)

Moreover, whilst public contestation has continued, it is important not to confuse the issue driving current conflicts. This is because although 'political antagonism may masquerade as devotion to preservation of natural resources' (Pillay, 1997: 7), as we shall see in Chapter 5, whilst for some communities their concern has been over environmental impacts, this has not been the case in all situations.

Knowledge: Finally, loss of trust has fed into interdependent knowledge conflicts. There have been attacks on the science and data around which the sustainability institution has been constructed and which have given it its meaning. Like their Scottish and French counterparts, Greek natural scientists have carried out many studies on the environmental impacts of farms, concluding that the area under the cages is stable (Kiaoudatos, 1998; Mantzaurakos et al., 2007). At times, these findings appear to conflict with community observations of the state of the lagoons in their area. Producers also have their own data as they take measurements for the EIAs but, as they recognise themselves, the public do not trust their data. This has led them to call for greater transparency and control by the state:

> It doesn't work always because practically people don't trust that the state does its job and they say ok you are monitored in theory, but in practice you are out of control and your farms are polluting because they are not controlled properly. So it's another issue to prove that you are monitored.
>
> (Interview: Greek seabass/seabream producer, 2012)

Q. The Ministry of Development said that this information is private.
Yes, well no it shouldn't be. It should be public. I think that it will be much better than people think it is. People think terrible things about aquaculture ... So I am very much in favour of transparency. In some cases [data] are not available. In some cases there is a lack of credibility. For example, we test four times a year the benthos; the water column and we publish as part of our ISO, as part of our commitment to our ISO to make this information publicly available. Anyone can come in and see them. But there is a lack of credibility. These are your results, we don't believe them. So I would love for the state to come, measure, publish, because then everyone will know what it is. If I have a problem, I want to know it. I am the first to want to know it. And if I don't, I also want to know it. I am a big believer in monitoring. Because if there is a problem. Usually there should not be a problem if the siting is correct and the EIA is respected. That's the first rule. If there is a problem for whatever reason then the farmer needs to know about it, because it is about economic sustainability. You will have a problem and no longer be economically viable. That is as far as the environmental aspect is concerned.

(Interview: Greek seabass/seabream producer, 2012)

Ecology–economy depoliticisation narratives are thus once more apparent in conflicts at the frontiers of interdependent knowledge types. For example, talking about the measurements in a situation of non-compliance with carrying capacity, one producer said this:

The main thing here is that the measurement is of the impact of specific farms and there the data were real, so the measurements are real.

(Interview: Greek seabass/seabream producer, 2012)

Table 4.3 Summary of political work over Greek seabass and seabream farm/environment interactions

Greek seabass and seabream			
Social constructions of sustainability	Dominant de-politicisations of ecology–economy interdependencies are frequently attacked but not fully replaced by alternative constructions of sustainability. Seeds of alternative ecosystem approaches increasingly brought within the institution via implementation of EU rules		
Frontier politics of polity interdependencies	<u>Territory</u> Greek state-wide approach dominates and no political work to re-assign authority to another administrative scale despite criticisms	<u>Regulation</u> Public rules dominate and non-compliance has taken precedence over private self-regulation; New public-private consultation to renew and legitimise public rules	<u>Knowledge</u> Different knowledge types produced (scientific, producers, state) but conflicts over credibility of these different knowledges; Conflicts over environmental impacts

Consequently, I cannot summarise, as I did in the case of Scottish salmon, that politicisation of nature–society interdependencies has been addressed through claims that impacts are known, knowable and monitored. Additionally, those actors and individuals who attack dominant knowledge claims have not so far positioned themselves as holders of alternate knowledge who could thereby contribute to bringing about a change in the contents of the sustainability institution.

Conclusions

In this chapter I have explored how actors in Scottish salmon, Aquitaine trout and Greek seabass and seabream have engaged in political work governing fish farm/environment interactions. As I summarised at the start of the chapter, research on observed environmental impacts of fish farming aquaculture has already set out what the issues related to farm/environment interactions 'could be'. These have included both organic and inorganic impacts as well as interactions with wild fish both at the localised farm scale and at wider ecosystem scales of the watershed. However, it is only through empirical research that we can grasp how actors in the case study sites have actually framed fish farm/environment interactions and chosen to govern them.

The application of the framework has allowed us to see multiple narratives which have been mobilised by different actors to define sustainability in terms of environmental impacts of farms and interactions between production and biophysical processes. Further, it has revealed the temporal, as well as spatial, scale of sustainability politics. Social constructions of sustainability of farm/environment interactions in each territory and sector have evolved over time and continue to be shaped today. Indeed, through studying political work on the sustainability institution, we can see that this process is never complete but has become part and parcel of what actors governing fish farming aquaculture do.

In this regard, there are important differences in the policy trajectories of sustainability between Scottish salmon, Aquitaine trout and Greek seabass and seabream to be grasped. First, in Scottish salmon, whilst salmon farm/environment interactions were initially represented as occurring at a local farm scale and were depoliticised, today this is no longer the case. On the one hand, according to a majority of actors' representations, farm/environment interactions are known to also occur within the wider water body and have been co-regulated through the gradual development of a range of public and private instruments also set at this scale. On the other hand, whereas organic and inorganic impacts tend to be considered as both known and manageable, and thresholds for their acceptable limits have been set in a raft of regulatory instruments, interactions between farmed and wild fish remain a controversial issue. This has involved actors in mitigation policies on siting, fallowing and in the collective commissioning and financing of research into alternatives (e.g. alternative approaches to replace chemicals in the treatment of sea lice).

Yet actors have only come to these changed conceptions of the problem through an accumulative political work in which narratives on farm/environment interactions have over time become entangled with those on changing regularities of political power. This has included political work separating out a Scottish governing approach as distinct from either a UK or an EU one,[71] including Scottish societal ownership of the problem; shifting regulatory authority first from private planning to public control and then through to co-regulation (public/private); and finally through to scientification of the controversies. Consequently, we can see that actors have clearly mobilised over the frontier politics of these policy interdependencies, creating contingencies to shift responsibility for these problems in Scotland. However, it does not follow that a 'Scottish' consensus has emerged over these questions. On the contrary, tensions persist between actors, as I discuss more fully in the book's concluding chapter.

By comparison, in regard to Aquitaine trout, initially trout farm/environment interactions were governed through state-wide general rules institutionalising anthropocentric ecosystem approaches. These already conceptualised organic, inorganic and fish disease impacts at the scale of the watershed. This scalar conceptualisation of the problem has not changed since. Indeed, actors have elaborated this construction of the problem to include other interactions between farmed and wild fish beyond the spread of disease. These include interactions between the physical infrastructure of the trout farms and aquatic ecological continuity, whereby the weirs in place to redirect water flow through the farm have been viewed as causing river fragmentation and hence impacting negatively both on ecological continuity[72] and (at times) on water flow. These recent institutionalisations of water quality have been supported by a (new) coalition of actors whose shared social representation of the rivers' future has potentially replaced former anthropocentric approaches with quasi-ecocentric ones. This has given rise to heated discussions around the continued compatibility of trout farming with the implementation of the WFD in France and uncertainty over the thresholds being negotiated.

In these debates, attempts have been made by trout producers to (re-)negotiate the sustainability institution. This has not only involved making changes to production practices, but also been achieved through political work. This has included vocalising alternative local territorial visions of rivers; developing sector-specific private self-regulation to demonstrate a 'clean environmental identity' as a counterpoint to dominant public rules; and generating self-monitoring data as potential leverage for new representations of impacts. In this way, actors have mobilised resources within the changing regularities of political power in conflicts over sustainability's very construction. For now, ambiguities persist over the extent to which 'sector' actors can re-shape the contents of sustainability governing farm/environment interactions or whether its contents will be determined by distant governors of water.

The governing of Greek seabass and seabream farm/environment interactions has also produced ambiguities. This is because seabass and seabream farm/environment interactions have been governed through dominant sustainability narratives depoliticising ecology–economy interdependencies which have not (yet) been replaced by alternative narratives. However, the government of farm/environment interactions has taken place in a context of legislative anarchy, including non-compliance with the rules. Consequently, over time, even if mitigation policies, e.g. on fallowing, have been put in place, the contents of the sustainability institution have been consistently attacked, especially by local communities.

However, alternative narratives to replace it have not yet been mobilised by new coalitions of actors. Instead, dominant narratives have been sustained through re-stabilising a Greek state-wide approach in line with EU policy, new public/private consultation to renew and legitimise public rules, and conflicts over the credibility of different knowledge types. This has resulted in a changing conceptualisation of the problem of farm/environment interactions from being a problem posed at the local farm scale to one posed instead at the scale of the water body, thus sowing the seeds for mobilisations around new narratives problematising ecosystem approaches to ecology–society interdependencies.

Following from this analysis of policy trajectories, we can summarise the comparative European aquaculture policy on fish farm/environment interactions. In Scotland, the EIA has been a key instrument at the centre of this policy, yet its role has extensively evolved over time. Whereas in the early years, the EIA was a tool for growth, it has increasingly been extended to becoming an ecosystem regulatory instrument grasping multiple impacts at different scales and in relation to biodiversity as well as organic and inorganic impacts. It has been supplemented by rules on fish health and on standards for the containment of fish as well as farm management agreements addressing issues such as sea lice. This shift has been accompanied by an opening up of the governing space of aquaculture to a range of public and private stakeholders whereby conflicts (with the exception of some e-NGOs) have been brought within governing arenas. An accumulation of tensions on sea lice treatments and escapees can be observed and these are likely to dominate conflicts into the future.

In Aquitaine, initial uses of EIAs were already set at wider ecosystem scales and have over time been extended to address questions of biodiversity and fish health. These have been added to by WFD norms on aquatic ecological continuity and low flow water targets for rivers. Whilst controversies over water policy in general have been brought within river basin committees, French trout farmers are on the margins of these decisional arenas. At present, no public governing space equivalent to the Scottish Ministerial working groups exists. Instead, private economic actors have created their own private instruments and public/private partnerships on the regional scale to compensate for any lack of public sectoral governing arrangements. Accumulating tensions relate to finding synergies between river water for fish farming and river water for migratory fish.

In Greece, the application of EIAs has also witnessed an expansion in their coverage towards regulation at a wider ecosystem scale and including fish health and biodiversity. However, these changes have been overshadowed by community attacks on aquaculture farms and regulatory challenges stemming from legislative anarchy and non-compliance with rules. Tensions accumulate around trust of the knowledge and data generated by EIA measurements. However, whereas there have been many attacks on the meaning of sustainability as win–win ecology–economy interdependencies, as yet no clear alliance of actors supporting an alternative sustainability has been forthcoming. New spatial planning instruments could potentially create new governing arrangements in which such alliances can be built.

Notes

1 These questions were raised within the third sustainability narrative: 'ecosystem approaches problematising ecology–society interdependencies' – see Chapter 2.

2 This literature is mainly within ecology and is far reaching – I cannot do justice to the richness of this literature in this chapter. I have based my summary on a handful of articles presenting research results specific to salmon farming, trout farming and seabass and seabream farming in Scotland, France and Greece.

3 This method does not close the door to the making of policy recommendations. Rather, through identifying specific obstacles to actors meeting their own goals, the method can enable research to produce targeted advice in situations of diversity, rather than universal policy solutions.

4 Scholars have also documented change in predator/prey relationships, see Wiber *et al.*, 2012.

5 A literature arguing for the adoption of ecosystem approaches for aquaculture has also argued for the development of multi-trophic aquaculture production practices to off-set impacts. For example, see Costa-Pierce, 2010; Ferreira *et al.*, 2014.

6 The Crown Estate is the UK Sovereign's public estate managed by Crown Estate Commissioners. It includes within its ownership the sea-bed and the foreshore. Revenues for any leases of this land go to the UK government.

7 The Scottish Office was the office of the UK Government in Scotland. Since 1999 and UK devolution it has been replaced with an elected Scottish Parliament and Scottish Government.

8 This directive was implemented through the Environmental Assessment (Salmon Farming in Marine Waters) Regulations 1988.

9 See, for example, Berry and Davison, 2001; Ross, 1997; Scottish Wildlife and Countryside Link, 1988.

10 Scottish Executive, 2003.

11 These new actors included the SG; SP; Marine Scotland; SG regulatory agencies; producers' organisations (e.g. Scottish Salmon Producers Organisation: SSPO); e-NGOs; and wild fisheries associations. Working groups also discussed other issues related to access to farm sites and marketing – see Chapters 5 and 6.

12 At the time, the SG was called the Scottish Executive (SE).

13 For the first two terms of office, the Scottish Government was called the Scottish Executive. The name was changed when the Scottish National Party was elected to a majority government in 2007.

14 This strategy covers all aquaculture production, including shellfisheries, in Scotland. In this book, I only refer to those elements of the strategy which concern salmon farming.

15 Interviewees were keen to stress that Scottish production accounts for 90 per cent of UK aquacultural value and volume – see also DEFRA, 2015.

16 The industry route has been to engage in the Federation of European Aquaculture Producers which is the European-wide federation representing fish farmers.

17 For example, policy officials within Marine Scotland, which is part of the Scottish Government and brings together scientific expertise on fish health issues.

18 This is a private label relating to animal welfare developed by the RSPCA.

19 Chemicals, medicines and disinfectants (antifoulants and pesticides) and the use of chemicals to control and prevent disease are covered *inter alia* by the Medicines Act 1968, the Environmental Assessment Regulations 1988 and, more recently, the EU Fish Health Directive 2006/88/EC.

20 EU Strategic Environmental Assessment Directive, 2001.

21 Consent (or not) is granted to applications concerning their discharges (e.g. feeds, faeces, size of farm, cages, layout and chemicals used).

22 The overseeing Ministerial working group set up in 2003 had already provided the impetus for the development of a Scottish Producers Code of Good Practice (interviews).

23 This Act contains a statutory requirement that companies sign Farm Management Agreements (FMAs).

24 Such as site optimalisation:

> whereby areas of farm where there are lots of little independent operators, operating on different cycles, have been rationalised by the industry with the government's help. So you have single companies now, occupying significant areas, synchronising their production all year – all fish in, all fish out – at the same time with mandatory fallowing period, and this is something which we very much welcomed. That has been progress.
>
> (Interview: Salmon fisheries boards' representative, 2012)

25 Studies have also been conducted on peoples' images of fish farm/environment interactions. These have concluded that: 'Only 1% of those surveyed in Scotland believed that fish farms were the main source of seawater pollution: only 4% believed it was a contributory source' (Whitmarsh and Wattage, discussing a survey carried out by Hinds *et al.*, [2002]; 2006: 109).

26 For example, the engineering standard mentioned emerged as an outcome from a research project commissioned by SARF to manage containment.

27 Available at: http://donstaniford.typepad.com/my-blog/2017/01/fish-farmageddon-scottish-salmon-loses-chemicals-arms-race-.html and reported in *The Sunday Times*: 'Salmon industry toxins soar by 1000 per cent' (1 January 2017).

28 Under the Environmental Code detailing authorisations for installations requiring environmental assessment (ICPE) 976–663.

29 Initially under the Water Law of 1992.

30 Original French: 'Il faut laisser l'eau dans la rivière pour les autres usages et pour la vie biologique derrière.... C'est même le droit civil, il faut laisser l'eau pour celui qui est en dessous quoi?! Sinon on fait "Jean de Fleurette"'.

31 For example, the river basin committees and water agencies acting within the water development and management plans (Schémas Directeur d'Aménagement et de Gestion des Eaux [SDAGEs]) and water management plans (Schémas d'Aménagement et de Gestion des Eaux [SAGEs]) and in consultation with the territorial collectivities.

32 This process could take up to eight months.

33 Original French: 'L'eau est à tout le monde. [Il y a une] responsabilité de respecter le bon état des cours d'eau avec tous les usagers.'

34 ANPER-TOS: Association Nationale de Protection des Eaux et Rivières – Truite, Ombre, Saumon – National association for the protection of waters and rivers – trout, grayling and salmon.

35 Both through the revision of the Environmental Code and the Water Law 2006 (loi sur l'eau et milieu aquatique [LEMA] 2006).

36 For example, within the Ministry of Ecology, water agencies, river basin committees and ONEMA (the French national water and aquatic environmental agency).

37 The main objective of the Water Framework Directive (WFD) was to achieve good ecological status of water in all Member States by 2015.

38 Orignal French: 'C'est un gros problème pour la vie biologique c'est que les cours d'eau ne sont pas en libre variation avec des zones de méandre, des zones de rapides, des zones de dépôt, une libre circulation des eaux, des sédiments qui fait que le milieu est vivant. Il est corseté par une succession de petits seuils qui fait qu'on est plus dans une succession de plans d'eau qu'un cours d'eau vivant.'

39 Within the Ministry of Agriculture.

40 Equivalent, for example, to the Scottish strategy.

41 Original French: 'Pour la mise en œuvre de la DCE on a raisonné de matière globale et non par activité.'

42 For example, ONEMA, the Water Agency, SAGEs.

43 GDSAA: Groupement de Défense Sanitaire Aquacole d'Aquitaine.

44 90 per cent of Aquitaine fish farms are certified as being 'safe'.

45 This was developed within the regional AREA programme. 'AQUAREA is the fish discharges, the environment of everything, that the site is clean, that all the constraints are implemented, economy of energy, sanitary aspects, all productive processes' (Interview: French trout producer, 2012).

46 Original French: 'En passant par AQUAREA … ça permet aux pisciculteurs d'avoir une information à titre individuel sur leurs élevages et en même temps d'avoir une communication auprès des administrations de façon collective pour être défendus.'

47 Original French: 'Dans le pays Basque on a toujours été "clean"'.

48 Original French: 'Dans des autres régions, il y une perception négative de la pisciculture, faite par des écologistes, mais peut-être ils [les pisciculteurs] travaillent mal aussi.'

49 Original French: 'Avant la truite était une sentinelle … Là où il y a une pisciculture l'alimentation doit être propre. S'il y a un gens qui met de la merde dans la rivière … mes poissons vont crever'.

50 Original French: 'Aller dans le sens de l'environnement … on le fait tous.… Moi aussi j'ai besoin de bonne qualité sinon mes poissons ils ne peuvent pas pousser'.

51 Original French: 'On est accro à l'environnement mais accro comme une drogue on en dépend complètement de l'amont et on dépend à l'aval car on est obligé d'avoir des rejets minimums pour qu'on nous donne l'autorisation de continuer à exploiter, donc nous l'environnement on vit dedans, on baigne dedans.'

52 CIPA: Comité Interprofessionnel des Produits de l'Aquaculture. This interprofessional body brings together trout producers, feed companies and processors at a French-wide scale.

53 Original French: 'Problème en amont – On est absolument non présent au niveau européen. Clairement.'

54 Original French: 'On a des textes européens et des indications très claires sur les obligations sur la restauration de ces cycles migratoires. Ça, c'est imposé par l'Europe.'

55 Environmental Code; Water Law 1964 – revised in 1992 to preserve water courses and in 2006 to meet WFD objectives by 2015; loi de pêche 1984; the loi Barnier 1995, plus associated Ministerial decrees, for example on discharges; and further commitments made by the Grenelle Environment programme in 2007.

56 For example, in high season when the water in the river is low, discharges from sites could be higher than those permitted as they will be more concentrated in the river. Yet on average – e.g. over a year – the same levels of discharges could be measured as being below those permitted.

57 For example, producers ask what happens if there is a drought: 'What do we do? Sell all our fish?'
58 Officials from the Ministry consider that requested efforts are not disproportionate, and also that there will be public funding and support to help trout farmers; trout farmers consider that the funding will not cover the full cost and are very concerned about uncertainties and finding themselves unwittingly in breach of the law at a time when they are working on their environmental impact. On this issue, the IUCN report (2011) also noted challenges due to a lack of knowledge and technical support to adapt to the rules.
59 Of course, producers do have regular exchanges with officials from the fish farming division of the Ministry of Agriculture, but thus far no sectoral strategy equivalent to the one institutionalised in Scotland has been forthcoming – see Chapter 5.
60 ONEMA is the French national water and aquatic environmental agency.
61 Original French: 'Les experts de l'ONEMA se rendant sur un cours d'eau sont capables d'en faire l'expertise par rapport à l'écosystème normal.'
62 Original French: 'Ils vivent dans un autre monde, ils ne sont pas comme nous.'
63 A lot of work has been undertaken to reduce emissions. For example: work on feeds; modernisation of technology; oxygenation, genetic selection; trout sterilisation (so that they do not produce in the wild if they escape); knowledge about stocking practices; etc. (IUCN, 2011).
64 IDAqua is an auto-evaluation tools which sets economic, ecological and socio-territorial indicators, enabling each fish farmer to establish an environmental and economic balance sheet of their farm (http://frenchfoodintheus.org/805; Tocqueville, 2012).
65 Original French: 'Mais en même temps, les administrations grâce au GDSAA obtiennent des infos dont aucune autre région ne dispose.'
66 Original French: 'Et ensuite au quotidien, ça demande une technicité importante pour maîtriser tous les paramètres de l'eau…. Ça demande une grande disponibilité, de la passion et une grande technicité; ils ne sont pas ingénieurs pour autant mais ils sont en permanence en train d'enregistrer des données pour continuer à progresser.'
67 Article 24 of the 1975 constitution.
68 Law 1650/86.
69 Responsibility for governing aquaculture in Greece has moved around different Greek national administrative and regional departments which have also been re-organised and re-named (interviews). According to the Food and Agriculture Organisation (FAO), licensing authorities that currently operate are the Ministry of Maritime Affairs, Islands and Fisheries/Directorate for Aquaculture and Inland Waters and the Ministry for the Environment, Energy and Climate Change. See: www.fao.org/fishery/legalframework/nalo_greece/en.
70 30 per cent higher in some cases (interviews).
71 Although, of course, containing elements of both.
72 As we have seen, by acting as obstacles to migratory fish heading for their spawning grounds.

References

Adams, T., Black, K., MacIntyre, C., MacIntyre, I., Dean, R. 2012. 'Connectivity modelling and network analysis of sea lice infection in Loch Fyne, west coast of Scotland', *Aquaculture Environment Interactions*, 3: 51–63.
Aubin, J. 2009. 'L'aquaculture pollue-t-elle les mers? Les 2èmes journées Recherche Filière Piscicole', *Journées de la Recherche Filière Piscicole* [online]. Available at: www.journees-de-la-recherche.org/JRFP/page-JRFP1024.php.

Bailey, C. 2008. 'Human dimensions of an ecosystem approach to aquaculture'. In Soto, D., Aguilar-Manjarrez, J. and Hishamunda, N. (eds) *Building an Ecosystem Approach to Aquaculture*. FAO/Universitat de les Illes Balears Experts Workshop. 7–11 May 2007, Palma de Mallorca, Spain. FAO Fisheries and Aquaculture Proceedings No. 14, Rome: Food and Agriculture Organisation (FAO), pp. 37–46.

Belias, C., Bikas, V., Dassenakis, M., Scoullos, M. 2003. 'Environmental impacts of coastal aquaculture in Eastern Mediterranean bays: the case of Astakos Gulf, Greece', *Environ Science and Pollution Research*, 10(5): 287–295.

Berry, C., Davison, A. 2001. *Bitter Harvest: A Call for Reform in Scottish Aquaculture*, London: World Wide Fund.

Bouleau, G. 2014. 'The co-production of science and waterscapes: the case of the Seine and the Rhône Rivers, France', *Geoforum*, 57 248–257.

Bouleau, G., Pont, D. 2015. 'Did you say reference conditions? Ecological and socio-economic perspectives on the European Water Framework Directive', *Environmental Science & Policy*, 47: 32–41.

Bouleau, G., Carter, C., Thomas, A., Boët, P., Salles, D., Auby, I., Oger Jeanneret, H. 2016. 'Suivre les médiations entre connaissances et décisions dans les dispositifs participatifs de gestion de l'eau: nouveau cadre pour comparer l'application de la DCSMM et de la DCE'. Paper presented at the conference 'Alter-Eau, au-delà des dispositifs institutionnels: quelles formes alternatives de participation à la démocratie de l'eau?', GEOLAB, University of Limoges, 16–18 November, Limoges.

Boy, D., Brugidou, M., Halpern, C., Lascoumes, P. 2012. *Le Grenelle de l'Environnement: Acteurs, Discours, Effets*, Paris: Armand Colin.

Conan, N. 2000. 'L'environnement dans la réglementation et la gestion des piscicultures en Grèce'. In Petit, J. (ed.) *Environnement et aquaculture, Tome II: Aspects juridiques et réglementaires*, Paris: INRA, pp. 266–274.

Conan, N., Prieur, L. 2000. 'Les piscicultures marines et l'environnement en Grèce'. In Petit, J. (ed.) *Environnement et aquaculture, Tome II: Aspects juridiques et réglementaires*, Paris: INRA, pp. 257–265.

Costa-Pierce, B. 2008. 'An ecosystem approach to marine aquaculture: a global review'. In Soto, D., Aguilar-Manjarrez. J. and Hishamunda, N. (eds). *Building an ecosystem approach to aquaculture*. FAO/Universitat de les Illes Balears Experts Workshop. 7–11 May 2007, Palma de Mallorca, Spain. FAO Fisheries and Aquaculture Proceedings No. 14, Rome: Food and Agriculture Organisation (FAO), pp. 81–115.

Costa-Pierce, B. 2010. 'Sustainable ecological aquaculture systems: the need for a new social contract for aquaculture development', *Marine Technology Society Journal*, 44(3): 88–112.

Cromey, C., Nickell, T., Black, K. 2002. 'DEPOMOD – modelling the deposition and biological effects of waste solids from marine cage farms', *Aquaculture*, 214: 211–239.

DEFRA (Department for Environment, Food and Rural Affairs). 2015. *United Kingdom Multiannual National Plan for the Development of Sustainable Aquaculture*. Available at: www.gov.uk/government/publications.

Douillet, A-C. 2003. 'Les élus ruraux face à la territorialisation de l'action publique', *Revue française de science politique*, 53: 583–606.

Fernandez, S., Debril, T. 2016 'Qualifier le manque d'eau et gouverner les conflits d'usage: le cas des débits d'objectif d'étiage (DOE) en Adour-Garonne', *Développement durable et territoires* [online], 7(3).

Ferreira, J., Saurel, C., Lencart e Silva, J., Nunes, J. Vazquez, F. 2014. 'Modelling of interactions between inshore and offshore aquaculture', *Aquaculture*, 426/427: 154–164.

Folke, C., Kautsky, N. 1989. 'The role of ecosystems for a sustainable development of aquaculture', *AMBIO: A Journal of the Human Environment*, 18(4): 234–243.

Fourmond, S. 2000. 'L'eau douce réglementée: quels droits et quelles conséquences pour les piscicultures?'. In Petit, J. (ed.) *Environnement et aquaculture, Tome II: Aspects juridiques et réglementaires*, Paris: INRA, pp. 19–22.

Fousekis, P., Lekakis, J. 1997. 'Greece's institutional response to sustainable development', *Environmental Politics*, 6(1): 131–152.

Frankic, A., Hershner, C. 2003. 'Sustainable aquaculture: developing the promise of aquaculture', *Aquaculture International*, 11: 517–530.

Georgakakis, D. 2012. 'Une Commission sous tension? La singulière différenciation des personnels administratifs et politiques de la Commission européenne'. In Georgakakis, D. (ed.) *Le Champ de l'Eurocratie: Une Sociologie Politique du Personnel de l'UE*, Paris: Economica, pp. 43–84.

Germaine, M-A., Barraud, R. 2013. 'Les rivières de l'ouest de la France sont-elles seulement des infrastructures naturelles? Les modèles de gestion à l'épreuve de la directive-cadre sur l'eau', *Natures Sciences Sociétés*, 4(21): 373–384.

Glenn, H., White, H. 2007. 'Legal traditions, environmental awareness, and a modern industry: comparative legal analysis and marine aquaculture', *Ocean Development and International Law*, 38: 71–99.

Gueneuc T., Tocqueville, A., Aubin J., Michel T., Michel G. 2010. *Les indicateurs de durabilité pour l'aquaculture*, Guide méthodologique IDAqua.

Hambrey, J., Edwards, P., Belton, B. 2008. 'An ecosystem approach to freshwater aquaculture: a global review'. In Soto, D., Aguilar-Manjarrez, J. and Hishamunda, N. (eds) *Building an Ecosystem Approach to Aquaculture*. FAO/Universitat de les Illes Balears Experts Workshop. 7–11 May 2007, Palma de Mallorca, Spain. FAO Fisheries and Aquaculture Proceedings No. 14, Rome: Food and Agriculture Organisation (FAO), pp. 117–221.

Hinds, K., Carmichael, K., Snowling, H. 2002. *Public Attitudes to the Environment in Scotland 2002*, Research Findings 24/2002, Scottish Executive: Edinburgh.

Ifremer. 2003. *Guide méthodologique pour l'élaboration des dossiers de demande d'autorisation d'Installations Classées pour la Protection de l'Environnement (ICPE) en matière de pisciculture marine pour la région Corse*, Rapport Scientifique Technique, R.S.T. DEL/PAC/04–05.

IUCN. 2011. *Guide pour le développement durable de l'aquaculture: Réflexions et recommandations pour la pisciculture de truites*. Gland, Switzerland and Paris, France: IUCN.

Jobert, B., Muller, M. 1987. *L'Etat en action*, Paris: UF.

Karka, L. 2011. 'Spatial planning for aquaculture: a special national framework for resolving local conflicts', 51st Congress of the European Regional Science Association: Special Session Territorial Governance, Rural Areas and Local Agro Food Systems.

Keeley, N., Cromey, C., Goodwin, E., Gibbs, M., Macleod, C. 2013. 'Predictive depositional modelling (DEPOMOD) of the interactive effect of current flow and resuspension on ecological impacts beneath salmon farms' *Aquaculture Environment Interactions*, 3: 275–291.

Lascoumes, P. 2011. 'Des acteurs aux prises avec le "Grenelle Environnement". Ni innovation politique, ni simulation démocratique, une approche pragmatique des travaux du Groupe V', *Participations* 1(1): 277–310.

Lloyd, M., Livingstone, L. 1991. 'Marine fish farming in Scotland: proprietorial behaviour and the public interest', *Journal of Rural Studies*, 7(3): 253–263.

Mantzavrakos, E., Kornaros, M., Lyberatos, G., Kaspirisa, P. 2007. 'Impacts of a marine fish farm in Argolikos Gulf (Greece) on the water column and the sediment', *Desalination* 210: 110–124.

Nautilus Consultants Ltd. 2002. *Submission to the Scottish Parliament Aquaculture Enquiry*. Available at www.nautilus-consultants.co.uk/sites/default/files/Submission%20to%20 Aqua%20Enquiry.pdf.

Papadakis, N. 2012. 'Réforme des administrations publique et locale: le cas de la Grèce', *Outre-terre*, 32: 36–369.

Papapanagiotou, E., Trilles, J. 2001. 'Cymothoid parasite *Ceratothoa parallela* inflicts great losses on cultured gilthead sea bream Sparus aurata in Greece', *Diseases of Aquatic Organisms*, 45: 237–239.

Papoutsoglou, S. 2000. 'Monitoring and regulation of marine aquaculture in Greece: Licensing/regulatory control and monitoring guidelines and procedures', *Journal of Applied Ichthyology*, 16: 167–171.

Pasquier, R. 2015. *French Regional Governance and Power in France; The Dynamics of Political Space*, Basingstoke: Palgrave Macmillan.

Peel, D., Lloyd, M. 2008. 'Governance and planning policy in the marine environment: regulating aquaculture in Scotland', *The Geographical Journal*, 174(4): 361–373.

Pert, C., Fryer, R., Cook, P., Kilburn, R., McBeath, S., McBeath, A., Matejusova, I., Urquhart, K., Weir, S., McCarthy, U., Collins, C., Amundrud, T., Bricknell, I. 2014. 'Using sentinel cages to estimate infestation pressure on salmonids from sea lice in Loch Shieldaig, Scotland', *Aquaculture Environment Interactions*, 5: 49–59.

Pillay, T. 1997. 'Economic and social dimensions of aquaculture management', *Aquaculture Economics & Management*, 1(1/2): 3–11.

Priyan, O., Smith, H. 1994. 'Environmental impact assessment as a management tool for the use of chemicals for fish farming in the UK', *Aquaculture International*, 2: 59–64.

Read, P., Fernandes, T. 2003. 'Management of environmental impacts of marine aquaculture in Europe', *Aquaculture*, 226: 139–163.

Read, P.A., Fernandes, T.F., Miller, K.L. 2001. 'The derivation of scientific guidelines for best environmental practice for the monitoring and regulation of marine aquaculture in Europe', *Journal of Applied Ichthyology*, 17(4): 146–152.

Ross, A. 1997. *Leaping in the Dark – A Review of the Environmental Impacts of Marine Salmon Farming in Scotland and Proposals for Change*. A report prepared for Scottish Wildlife and Countryside Link, Perth.

Russell, M., Robinson, C., Walsham, P., Webster, L., Moffat, C. 2011. 'Persistent organic pollutants and trace metals in sediments close to Scottish marine fish farms', *Aquaculture*, 319: 262–271.

Saliou, V. 2010. 'Making Brittany a space for maritime politics: building capacity through the politicization of regional identity', *Regional & Federal Studies*, 20: 409–424.

Scottish Executive. 2003. *A Strategic Framework for Scottish Aquaculture*. Edinburgh: Scottish Executive.

Scottish Government. 2009. *A Fresh Start: The Renewed Strategic Framework for Scottish Aquaculture*, Edinburgh: Scottish Government.

Scottish Wildlife and Countryside Link (SWCL). 1988. *Marine Fish Farming in Scotland*. Perth: SWCL.

Soto, D., Aguilar-Manjarrez, J., Brugère, C., Angel, D., Bailey, C., Black, K., Edwards, P., Costa-Pierce, B., Chopin, T., Deudero, S., Freeman, S., Hambrey, J., Hishamunda, N., Knowler, D., Silvert, W., Marba, N., Mathe, S., Norambuena, R., Simard, F., Tett, P., Troell, M., Wainberg, A. 2008. 'Applying an ecosystem-based approach to aquaculture: principles, scales, and some management measures'. In Soto, D., Aguilar-Manjarrez, J. and Hishamunda, N. (eds) *Building an Ecosystem Approach to Aquaculture*, FAO/Universitat de les Illes Balears Experts Workshop. 7–11 May 2007, Palma de Mallorca, Spain. FAO Fisheries and Aquaculture Proceedings No. 14, Rome: Food and Agriculture Organisation (FAO), pp. 15–35.

Sutherland, M., Lane, D., Zhao, Y., Michalowski, W. 2009. 'A spatial model for estimating cumulative effects at aquaculture sites', *Aquaculture Economics & Management*, 13(4): 294–311.

Tanguy, H. 2008. Rapport final de la mission sur le développement de l'aquaculture. Ministère de l'Agriculture et de la Pêches. Ministère de l'Ecologie, de l'Energie, du Développement Durable et de l'Aménagement du Territoire. Available at: http://archives.agriculture.gouv.fr/publications/rapports/mission-sur/downloadFile/FichierAttache_1_f0/Rapport_H_Tanguy.pdf.

Tocqueville, A. 2012. 'Évaluer et améliorer la durabilité de la filière piscicole française: programmes IDAqua et ProPre'. Paper presented at the conference 'Towards Competitive and Sustainable Fish Farming', organised by INRA, CIPA, Itavi and Sysaaf, 27 February, Paris.

Whitmarsh, P., Wattage, P. 2006. 'Public attitudes towards the environmental impact of salmon aquaculture in Scotland', *European Environment,* 16: 108–121.

Wiber, M., Young, S., Wilson, L. 2012. 'Impact of aquaculture on commercial fisheries: fishermen's local ecological knowledge', *Human Ecology*, 40: 29–40.

Young, J., Brugere, C., Muir, J. 1999. 'Green grow the fishes-oh? Environmental attributes in marketing aquaculture products', *Aquaculture Economics & Management*, 3(1): 7–17.

5 Sustainability interdependence and access to fish farm sites

Environmental landscape aesthetics and coastal/rural development

Introduction

In this second empirical chapter, I examine the political and institutional work of public and collective private actors in the institutionalisation of sustainability governing access to fish farm sites in Scottish salmon, Aquitaine trout and Greek seabream and seabass. Whereas Chapter 4 (fish farm/ environment interactions) covered sustainability issues which can potentially be cast by actors as problems from the farm scale and up to the water body/ watershed scale, this chapter covers sustainability issues potentially defined as problems extending from the water body/watershed scale and up to the community or regional scale.

As I suggested in the introductory chapter, the early growth of fish farming aquaculture in Scotland, Aquitaine and Greece was directly associated with contributing to the 'social uplift of rural communities' (Pillay 1997: 4). Fish farming was seen as contributing to economic development in deprived areas and creating employment in rural and coastal communities (Rana, 2007). However, following its intensive growth, coupled with increasing doubts about its environmental effects and community conflicts over local development projects, an important question emerged over its continued development. This, it was argued, would only be achieved through its integration into the social and cultural fabric of local communities.

Over time, however, given local community hostility to fish farming in some areas, whether aquaculture indeed had a place in future coastal and rural development paradigms became less certain.[1] Could this industry be considered as part of local (and even regional or national) responses to current societal challenges? Or, as oppositional movements began to argue, should communities simply 'say no to farmed [fish]'.[2] Importantly, such debates on fish farming coincided with the growth of the tourism industry, often directly competing for access to space (Hall et al., 2013). Consequently, whereas in Chapter 4 the sustainable growth potential of aquaculture concerning fish farm/environment interactions was posed in terms of the carrying capacity of the ecosystem, this chapter raises the question of the sustainable growth of the industry in terms of its insertion in the local community and territorial development. This includes tackling sustainability from the perspective of environmental landscape effects (Young *et al.*, 1999) often not addressed in sustainability assessments (Maréchal and Spanu, 2010).

In this chapter, I first briefly set out what the issue of access to fish farm sites 'could be' as discussed in the scientific literature on fish farming aquaculture. I then proceed to examine in each sector and territory how access to fish farm sites has emerged as a contested issue in Scotland, Aquitaine and Greece. Applying my institutionalist sustainability interdependence framework, in each case, I will show how actor political struggles over nature–society interdependencies combined with struggles over the frontier politics of interdependence and territory, regulation and knowledge have determined the choices leading to different 'settlements' of sustainability interdependence governing access to farm sites. As in the case of Chapter 4, I conclude by comparing the different approaches and drawing out their respective similarities and differences in terms of aquaculture policy.

Potential issues in accessing fish farm sites: the story so far

There are a number of potential issues discussed in the literature on aquaculture concerning access to farm sites. Overall, scholars have come at this question from three main angles: (i) re-defining growth as a problem for aquaculture; (ii); proposing management tools to 'better integrate' fish farming into coastal development; and (iii) developing off-shore aquaculture.

First, aquaculture has often been presented as a growth industry. However, scholars working within ecosystem approaches analysing this industry have argued that this representation of aquaculture is not only misleading but also acts as an obstacle to debate over aquaculture's future role in contributing to food security (Costa-Pierce, 2010). This is because in a global context of declining wild fish stocks it is often argued that consumer demands for fish are increasingly being met through aquaculture, whose global output (it is further argued) has doubled in the past ten years (Culver and Castle, 2008: 93). Frequently, representations in official public documents define aquaculture on the worldwide scale as the 'fastest growing food production sector' (European Commission, 2009a: 2), with global growth rates averaging 6–8 per cent per annum and providing 'half of the world fish supply for human consumption' (European Commission, 2009a: 2). However, Costa-Pierce (2010) has argued that a closer inspection of growth figures reveals that if one removes China from the calculation, there is in fact very little global growth:

> The world is not eating half of its seafood from aquaculture. The world has watched, and is watching, a blue revolution ... in China. In 2006, China accounted for 67% of all global aquaculture production. ... Chinese aquaculture production is largely feeding China, not the world.
> (Costa-Pierce 2010: 89)

Indeed, within EU production, growth has slowed overall and is in decline for some species. For example, Ernst & Young found a −22 per cent EU growth rate between 2006 and 2008 (2008: 32). Seeing growth as a problem and not a feature of this industry, it is argued, enables the shifting of the debate towards how to govern an industry with negative growth, especially when consumer demand for seafood is high (in the EU, 60–65 per cent of consumer demand for seafood is met through imports; European Commission, 2009b: 10). Moreover, this directly raises the question of the place of fish farming aquaculture in society's future. For many scholars working on this industry, the answer is clear. For example, in his consideration of the urgent issue of food security in the Anthropocene, Costa-Pierce argued that 'aquaculture is one of the planet's best choices' (2010: 107). Consequently, seeing the problem as one of growth enabled him to argue that whereas aquaculture *should* be growing, it is not.

Reframing the problem as one of (lack of) growth leads to the second way in which the literature on ecosystem approaches to aquaculture has discussed the issue of access to farm sites. This is because, whereas it has been argued that 'growth has declined partly because of stringent regulation' (Nunes *et al.*, 2011: 369), removing this (often environmental) regulation is not a policy strategy that is often advocated. Instead, scholars have identified local conflicts over fish farming aquaculture and the lack of access to new farm sites as critical issues to be addressed through management tools. As Pillay argues, whereas scholars perceive that

> adverse environmental effects can be corrected … the hardest challenge is balancing economic, environmental and social costs.
>
> (1997: 11)

This has given rise to discussions about possible solutions for addressing local conflicts. Within sustainability narratives of 'ecosystem approaches problematising dynamic ecology–society interdependencies' (as discussed in Chapter 2), scholars have advocated going beyond EIAs as regulatory instruments to additionally mobilise wider planning tools such as 'Integrated Coastal Zone Management' (ICZM)[3] to seek to integrate discussions about aquaculture's future within inter-sectoral discussions on coastal futures (Soto *et al.*, 2008: 19). Put more directly, in the face of aquaculture decline, scholars have expressly argued for a change in the development paradigm *towards* an Ecosystems Approach (EA) to aquaculture, which includes for them some form of integrated planning, accompanied by community stewardship practices.[4] This would include the construction of a narrative about aquaculture which would link this industry to its recent past and

> develop the background, baselines, and place-based ecological and social contexts of aquaculture so that more informed decisions can be made by politicians, investors, and communities.
>
> (Costa-Pierce, 2010: 92)

In terms of building a 'social contract' for aquaculture (Costa-Pierce, 2010), this would include work to ensure that aquaculture's benefits are shared locally,[5] understanding that both job creation and infrastructure development brought about through fish farming can be factors of social resilience (Bailey, 2008). Scholars working on European aquaculture also consider that both the EU WFD (2000) and the EU Marine Strategy Framework Directive (MSFD; 2008) already embody the core principles of an EA and that both the legal instruments and social contracts for an EA already exist (Nunes *et al.*, 2011). Indeed, scholars have further recommended use of the EU WFD and ICZM both as policy initiatives for European planning (Read *et al.*, 2001), and for earning respect for aquaculture, understanding it to be a 'new' industry (Stead *et al.*, 2002).

Other possible solutions which have been proposed promoting 'place-based' approaches to aquaculture development have been made by researchers working on science–policy interdependencies and aim to encourage citizen engagement in knowledge generation. Grant (2010), for example, argued that in Canada's Bay of Fundy, where consultation and participation of coastal communities had not originally been forthcoming, new projects to develop participatory research and citizen science on the environmental impacts of aquaculture had had positive returns in regard to community 'buy-in' to aquaculture (Grant, 2010). Additionally, social scientists and lawyers working within a literature on planning have argued that, if fish farming aquaculture planning practices embrace core principles of good governance and modernisation projects which acknowledge changing state–market–civil society relations, then this can provide a way forward towards more open and democratic approaches to making local choices (Peel and Lloyd, 2008). However, scholars also point to a problem of diversity of regulation and contradictory legislation in a European context which creates incoherence and limits to legal implementation (Bermúdez, 2008; Peel and Lloyd, 2008). In this context, Bermúdez questions whether an ecosystem approach (EA) can be implemented within general legislation or whether aquaculture requires its own specific policy framework (2008).

Within all these scientific discussions on management tools, a major tension can be observed. On the one hand, researchers within ecological ecosystem approaches to aquaculture are clearly advocating management tools, e.g. ICZM, with a specific objective in mind: namely, that this will provide the means for communities to 'learn' that aquaculture is a positive response to societal challenges. On the other hand, researchers within social science are more concerned with transparent and democratic planning, whatever the outcome for aquaculture. Their objective is that any conflicts arising over what Young *et al.* have described as the 'privatisation of a common resource' (1999: 13) are managed in a transparent and democratic way. For them, use of ICZM does not foreclose options. Through ICZM, aquaculture's place in local communities may be confirmed – or not, if alternative local economies are instead desired by local actors.

Finally, a third approach to access to farm sites has been developed within a literature promoting technological innovation. Faced with the problem of lack of space in coastal waters for new farm sites, the argument is that aquaculture can move off-shore. For example, this would entail the design of new, larger cages, which could for example be attached to floating wind energy platforms off-shore in another version of sectoral integration:

> You cannot grow any fish in the south part of France now, in Corsica, no way. It is not because it is full, but because nobody wants to see it. But the idea of the wind farm is to go far away you cannot see anything … except if you are sailing.
> (Interview: European Commission DG Research public official, 2009)

Moving off-shore is argued to provide positive gains. According to Shainee *et al.*, these include less polluted water; natural dispersion and dilution of waste; deeper and bigger net cages; better fish growth; less fat and less disease; and the reduction of competition for space (2013: 135). Yet they also acknowledge challenges for technological innovation, including for cage design for biosecurity and the weather (Shainee *et al.*, 2013). Others argue that a number of knowledge gaps require to be addressed on potential environmental impacts, e.g. benthic biodiversity; escapees; carbon footprint; ecological footprint (feed); and the possible attraction of large predators (Holmer, 2010). There are also scientific challenges for monitoring off-shore farm/environmental interactions and the need for new dynamic models to grasp the interactions of off-shore and in-shore aquaculture through a system-scale approach rather than piecemeal planning (Ferreira *et al.*, 2014). Finally, scholars have argued that even when moving 'off-shore', it is unlikely that conflict-free areas exist (Chang *et al.*, 2014).

In summary, a number of potential issues surrounding access to fish farm sites are discussed in a varied set of literatures. These include the place for fish farming aquaculture within a society's response to the range of challenges with which it is confronted; managing community conflicts over the growth of this industry; and technological and governing challenges covering off-shore aquaculture. What is missing, however, is a political science approach comparing how actors in different territories and sectors have themselves defined the problem and rendered it governable. To begin filling this gap, in the rest of this chapter, I examine how access to fish farm sites has actually been framed through the political and institutional work of actors over Scottish salmon, Aquitaine trout and Greek seabass and seabream.

Scotland and governing access to fish farm sites

Overview of the changing sustainability narratives

Central tensions in the Scottish government around access to fish farm sites have emerged in response to an increasingly ambitious growth policy. These tensions are both over whether salmon farming *should* be growing, as well as over compensation measures for continued expansion of the industry. The initial growth of the salmon farming industry in Scotland followed a reduced role for public government through the granting of leases to companies by Crown Estate Commissioners[6] seeking to grow the industry in rural and coastal areas (Coull, 1988).[7] In the early years, salmon farming was widely welcomed.[8] Access to fish farms sites was governed, stressing salmon farming's potential socio-economic contribution and depoliticising ecology–economy interdependencies:

> It has created valuable employment opportunities, attracted inward investment and encouraged the participation of local people. These benefits are clearly of wider social and community significance in this relatively remote region.
>
> (Lloyd and Livingstone, 1991: 254)

These socio-economic benefits of salmon farming notwithstanding, questions were soon raised by different stakeholders over the processes leading to coastal planning choices. On the one hand, and as we have seen in Chapter 4, e-NGOs challenged the depoliticised ecology–economy sustainability institution governing Crown Estate monitoring of environmental impacts. On the other hand, different stakeholders, e.g. local authorities and regional councils, were concerned about the private nature of this planning process and their lack of control over choices concerning the sustainable development of rural economies.

In response, Crown Estate Commissioners drew up a comprehensive list of appropriate consultees in 1986.[9] These developments notwithstanding, this private management planning approach increasingly came under attack. A central criticism was that consultees ultimately remained marginal in choices over location decisions due to the 'proprietorial interest' exercised by Commissioners managing and developing the Crown Estate (Lloyd and Livingstone, 1991: 253). In particular, it was argued that the planning system did not allow for alternative choices to be considered by local communities and especially in the absence of an overall Scottish planning policy.

As documented in Chapter 4, criticisms of both 'win–win' ecological-economic arguments and private planning were jointly addressed by the Scottish Government (SG) post-devolution through the re-institutionalisation of the sustainability institution governing salmon farming. Now applying an anthropocentric ecosystem approach problematising dynamic ecology–society interdependencies, the SG granted local authorities the final say in decisions on fish farming applications and improved the role for regulatory bodies advising on the impacts of siting. These changes opened up institutional space for new arguments on access to fish farm sites to be advanced.

Up until now, actors contended, the environmental impacts of salmon farming had predominantly focused on biological impacts as distinct from the 'intrinsic value' of the site for other visitors and users (Muir *et al.*, 1999). This mattered because, although salmon farming had positive effects on employment, it was located in 'sensitive areas of natural beauty' (Muir *et al.*, 1999: 50). For example, Scottish Natural Heritage (SNH) argued both that the attractiveness of Scottish coasts potentially provided other benefits, e.g. for tourism, and that diversification of local economies was necessary (Scottish Natural Heritage, 2002). These arguments launched a broader discussion of environmental issues which problematised landscape aesthetics[10] and led to enhanced analyses of siting impacts.[11]

However, in a final twist to the tale over sustainability's trajectory in this regard, whereas developments in the 2000s had consolidated a particular anthropocentric ecosystem approach with emphasis on territorial insertion and stronger environmental control over development, the political strategy of the Scottish National Party (SNP) SG, which took up office in 2007, shifted the debate on access to fish farm sites once again – this time in favour of an explicit growth plan in a period of economic recession.[12] Described on interview by producers as the 'best supporting government in living memory of the industry', the SG began reforming planning processes to ease new applications for farm sites. This introduced new tensions into the sustainability institution, especially between the SG and producers versus regulatory agencies and NGOs, this time over the hierarchy of ecosystem services being endorsed.[13] In particular, questions were raised as to whether the governance of interacting ecosystem services should be handled through 'robust environmental intervention' or left to the 'free market system' mechanism (Interview: Scottish regulatory agency official, 2010).

Unpacking interdependencies

The sustainability institution governing access to salmon farm sites has been extensively worked upon by a range of Scottish actors mobilising simultaneously around the frontier politics of interdependence of territory, regulation and knowledge.

Territory: As we saw in Chapter 4, UK devolution in 1999 provided Scottish actors with new territorial resources and identities with which to govern and which were mobilised by them towards the development of a new sectoral strategy for aquaculture. The uptake of devolution's new resources also occurred in the case of planning. Following on the back of extensive criticism of the UK centralist (and private) nature of coastal planning, different stakeholders in Scotland expected that this would be territorialised post-devolution and in line with devolution's norms of inclusiveness and transparency (Peel and Lloyd, 2008). Described by Peel and Lloyd (2008) as indicative of a shift towards 'modern governance in Scotland', the SP[14] swiftly voted to extend the powers of local authorities authorising aquacultural planning decisions.

In the case of aquaculture, this re-assignment of authority meant that local authorities (and local communities) now had the final say in decisions over whether or not to integrate salmon farming in local coastal development futures. Very swiftly, conflicts emerged between some local communities and the SG over this very issue. Already a 'social contract' had been established that the North and East coasts of Scotland were 'closed off' to salmon farming to minimise interactions with wild salmon due to the presence of salmon runs. In addition, by the mid-2000s there were very few available sites, the best ones having already been filled. However, there were still possibilities for expansion: for example, through the reconversion of disused sites or through expanding into more exposed areas in the Outer Hebrides. However, not all communities were in favour of continuing development along these lines. Conflicts at the frontiers of interdependence and territory began to emerge over how different groups of actors politically aligned visions of territorial futures with representations of the environment (i.e. biophysical processes and landscape aesthetics).

On the one hand, since 2009, the SG has advocated a strong growth policy which has been presented as part of Scotland's future:

> Sustainable economic growth as a fundamental requirement of *our* nation (emphasis added).
> We are one of a handful of countries with the climatic and hydrographic conditions to farm salmon. Our pristine waters make our seas and coasts an ideal location for growing finfish, shellfish and seaweeds.
>
> (Scotland Food and Drink, 2016: 2)

> Aquaculture is a *nationally* important industry for Scotland, in particular the West coast and the islands where many communities depend on the employment and revenue it provides.
>
> (Scottish Government, 2009: 6; emphasis added)

> [T]he salmon sector in Scotland which is worth over 400 million pounds a year to the Scottish economy, which is great and that is a growing figure, although it's not massive. But what is important about salmon farming and aquaculture more generally is where it is in Scotland – in the coasts and a large number of rural communities.
>
> (Interview: Scottish public actor, 2010)[15]

Many local communities share this vision of Scotland's future in respect of their local territory. This has particularly been the case for some of the communities in less well-off parts of Scotland. For example, the socio-economic importance of salmon farming has been stressed in public surveys on local territorial differences regarding the desire (or not) for expansion of the industry, where expansion was most popular in poorer areas (Whitmarsh and Palmieri, 2009: 456).

However, some local areas are not in favour of fish farming and do not share this vision of their future. For some, this is because they are developing alternative economies, e.g. in Orkney where they have invested in renewable energy. However, within the salmon sector itself, a major challenge for the sustainability institution has arisen when alliances of actors have emerged within local communities around ecological arguments concerning landscape aesthetics:

> The requirement for the industry is for sheltered water which puts it into a very scenic part of the Scottish West coast. It is not as if the industry is trying to get into St. Andrew's bay or the Firth of Forth. It only wants to be in sea lochs, very scenic bits of the West coast so that brings it into contact with various landscape considerations.
>
> (Interview: Scottish public actor, 2012)

This is because arguments on environmental landscape aesthetics have often been evoked to make the case for a more ecocentric interpretation of the ecosystem approach which problematises ecology–society interdependencies. Further, such arguments have often been supported by a particular alliance of stakeholders, including middle-class second-home owners with potentially strong political influence over Local Authorities:

> We do supply employment in these remote fragile areas, but not everyone wants a fish farm. That might be because you have certain parts of the West coast that attract people who want to retire there or put up second homes. Vociferous people – and rich – and can afford to put up an argument against you.
>
> (Interview: Scottish salmon producer, 2010)

An awareness of conflicts within and between local territories over salmon farming's future has resulted in political work by both salmon farmers and the SG along four fronts across inter-territorial interdependencies. First, they have evoked EU strategies of growth to legitimise their own visions for coasts. Second, they have worked to insert themselves in coastal communities:

> Aquaculture takes place in the periphery of Scotland, in the regions away from the centre, it takes place in Orkney, Shetland, the Western Isles, the Highlands and Islands basically. Because that is quite a remote community, it needs community buy-in and involvement and we have to work within those communities and we need strong support from those communities, particularly from the political element in those communities and I think personally that works well if you have a good strong set of relations at a local or regional level.
>
> (Interview: Salmon producers' association, 2012)

Sustainability is with the communities as I mentioned … [we] need to be seen as a responsible business, that is providing jobs and there for the long term. We have been farming since 1965 and in a bigger way since the 1980s. We have been a feature in the Scottish communities for a lot of years and we want that to continue. We do a lot for the local community, apart from jobs, for example we sponsor a shinty team. There is no return … it is about supporting communities.

Q. Like corporate social responsibility?

Yes.

(Interview: Scottish salmon producer, 2010)

Third, companies have developed in-house policies to seek applications only 'where they are welcome'; and fourth, they have developed applications for more exposed sites. This political work has not been without its challengers. For example, radical e-NGOs, which are against salmon farming, have politicised the issue of only going 'where they are welcome', for example, through publishing emails between company workers in which they are seen to be disparaging of coastal communities. Moving towards more exposed areas has not been without difficulties either, generating a whole new set of considerations, particularly on staffing issues.

Consequently, once a separation of a Scottish approach to planning had taken effect, political work thereafter, led by the SG and producers, was directed towards minimising conflict within Scotland through developing a narrative which linked salmon farming to Scotland's future – a narrative which, nonetheless, has frequently been attacked by different groups of stakeholders espousing alternative visions of the social construction of post-devolution Scotland.[16]

Regulation: Political work managing conflicts at the frontiers of interdependence and territory have been interwoven with political work at the interface of private and public regulation. As I have already said, at the start of the 2000s, broad political support emerged in Scotland not only to territorialise the planning process, but also to democratise it.[17] As I have argued, this was accompanied in the case of salmon farming by a strengthening of the powers of regulatory agencies advising on planning, leading to the implementation of an anthropocentric ecosystem approach problematising ecology–society interdependencies. This new approach, following on from the (failed) sustainability institution depoliticising ecology–economy interdependencies, was stabilised in all the institutional work at this time on environmental landscape aesthetics.

In the mid-2000s, arguments which were mooted first by the SNH, and supported by both the SG at the time and other stakeholders, led to them proposing the introduction of a new environmental concept of 'visual carrying capacity' into planning processes. To turn this concept into a regulatory tool, SNH created a new type of environmental impact assessment, namely a 'visual impact assessment'. The visual impact assessment was coherent with an environmentally and territorially focused ecosystem approach to growth, setting out guidance for producers on how the design of their farm sites could 'fit with nature' (Costa-Pierce, 2010: 108):

> We try to give guidance to say if you are going to put in long lines, this is a mussel farm, rather than a fish farm, they are better if they run along the shore than across the shore for example. And tucking things into bays can make it less visually intrusive than sticking it out in the middle and so on. We provide guidance on how to fit developments into the character of the landscape and to do with scale.
>
> (Interview: Scottish regulatory agency official, 2010)

Whereas the newly negotiated sustainability institution was being instrumentalised along these lines, the growth policy launched by the (SNP) SG in 2009 began to create tensions at the heart of the institution. In a situation of economic recession, a number of problems began to be identified within the planning system. Many of these problems were defined as such during the discussions in the working group on planning. Producers, for example, complained that the licensing process was too cumbersome, and unwieldy and therefore ineffective for meeting growth targets. In response, the SG initiated a reform of the planning system,[18] issuing instructions to its regulatory agencies to streamline their advice processes (e.g. no longer asking for duplicate information). This included changing the methodology of the process and engaging in discussions with producers upstream of applications being submitted:

> Obviously as I said early on, if a developer says, look, I've worked out all my plans, here are my plans, and you just react to it, that brings us into a very confrontational situation very often. Whereas if we can have talked to them at an early stage and said look these are our difficulties and these are our constraints and there must be a win–win in here somewhere. So our business is not to constrain the industry but to guide them to bring forward a development of a kind and in a location and managed in such and such a way that you can get your objectives and we can achieve ours. So this kind of guidance and advice that we give is related to that. At the level of an individual development this is about the layout of your cages in relation to the specific layout of the landscape, but that is not the whole of it.
>
> (Interview: Scottish regulatory agency official, 2010)

This very growth-focused policy has been cautiously received by e-NGOs and other stakeholders, such as the wild salmon fisheries' boards. For example, whereas e-NGOs have argued that they support expansion, they were keen to make sure that 'it happens in the right way' (Interview: Scottish e-NGO,

2010). A related concern has been that whereas the sectoral plan and its associated working groups had initially been constructed within a spirit of participatory democracy around concepts of nature providing multiple benefits to Scottish society, the style of public/private relations more recently being institutionalised has appeared de-regulatory. As a result, the plan appears to prioritise nature in terms of benefits for salmon farming – thus potentially opening up space for further policy capture by strong economic interests, such as those held by producers.

Knowledge: Finally, as was the case for fish farm/environment interactions, actors have managed the different tensions within the sustainability institution through the generation of new knowledge and a 'scientification of controversies' (Weingart, 1999). For example, whereas the attractiveness of coasts can be a contested concept, visual impact assessments were not developed through discussions with local communities but through the mobilisation of landscape advisors and scientists:

> It is a science really … a sort of architectural science – not my field.
> (Interview: Scottish regulatory agency official, 2011)

> Well it is peoples' impression that it is subjective, that what one person sees as an impact, another person might not, but in terms of actual expert judgement, you can have a definitive answer. People struggle with this – I struggle with it – but if you speak to a landscape designer or architect – industry obviously struggle with it too. We realise that and we are planning some further training and workshops. Because industry like to do a lot of the assessments themselves. But I think in terms of landscape, whether that is feasible or not – it is quite a specialist skill.
> (Interview: Scottish regulatory agency official, 2011)

Additionally, on questions of conflicts between tourism and salmon farming growth, producers working either on their own or through SARF (the Scottish Aquaculture Research Forum) have commissioned studies to 'disprove' claims that aquaculture is 'bad' for tourism. In this research, it was found not to be an issue. This research has therefore been mobilised by actors in response to arguments advocating more ecocentric applications of ecosystem approaches.

Finally, although a consensus was observed that moving to off-shore/ exposed sites is where the industry should go, so far no-one has done so. Individual company initiatives have developed technology for exposed and off-shore sites, such as multi-skin submergible cages, but for the moment, this technology remains untried and has not been turned into a commercial operation. Business plans for growth are all still premised upon surface technology.

Table 5.1 Summary of political work over Scottish salmon access to farm sites

Scottish salmon			
Social constructions of sustainability	De-politicised ecology–economy interdependencies are replaced with anthropocentric ecosystem approaches. However, tensions emerge over hierarchy of ecosystem services being institutionalised		
Frontier politics of polity interdependencies	<u>Territory</u> Separation of Scottish approach; Political work by producers and SG to reduce conflicts over how different actors have connected visions of territorial futures with representations of the environment	<u>Regulation</u> From private planning to public control; Tensions over regulation as supporting nature and providing multiple benefits to society versus supporting nature and providing benefits to private economic actors	<u>Knowledge</u> Scientification of controversies in re-institutionalisation

Aquitaine and governing access to fish farm sites

Overview of the changing sustainability narratives

The main thread running through the story about the government of access to farm sites in Aquitaine has been one of industrial decline. Tensions have emerged over what is widely perceived as a governmental policy of control of fish farming activities rather than a policy of growth. Today's situation stands in contrast to the early years. This is because in the 1970s, trout farming in Aquitaine was considered to hold promise for the economic development of rural areas and territories (Pyrenees, Basque country, les Landes). For example, in Les Landes, trout farming offered a new solution to revitalise the area following the collapse of the pine resin industry.[19] Additionally, by addressing hunger and food security (IUCN, 2011), it carried the political support of the Regional Council at this time. However, emerging tensions in the sustainability institution governing trout farming in the mid-2000s between anthropocentric and quasi-ecocentric ecosystem approaches (as analysed in Chapter 4) resulted in the redefining of growth as a 'problem'. Indeed, the tension between different actors' interpretations of French rivers' futures as 'living' versus 'wild' rivers (Germaine and Barraud, 2013) became acute in actor mobilisations over access to fish farm sites. This is because even though Aquitaine has remained the first French region for trout production, in fact between 1997 and 2007 production declined by 19 per cent (Agreste, 2011).[20]

Whereas this decline in growth could be attributed to a number of factors including marketing strategies (see Chapter 6), the ambiguity of the French government's response to its decline fuelled arguments that a central problem was the way in which sustainability was being defined. As argued by producers, if the issue was one of sustainability (as defined by them), then they would have no dispute with the administration. This was because, according to them, sustainability had an economic dimension to it as well as a social and ecological one, wherein an economically alive industry was central. However, they argued, this was not the dominant interpretation:

> We are working with a specific administrative environmentalist vision of action … we have environmental regulation which is not regulation on sustainability. If today we applied the law strictly we would close half of the fish farms in France.
>
> (Interview: French trout producer, 2012)[21]

> The question is, do we put humans in the environment or do we exclude them from the environment? There is a tendency here to exclude humans from the environment.
>
> (Interview: French trout producer, 2012)[22]

In alliance with officials from the fish farming division of the Ministry of Agriculture, producers' associations have consequently worked to redefine lack of growth as a public problem for Aquitaine (and for France). In so doing, they have tried to reshape the contents of the sustainability institution, arguing that sustainability *should* include a concern over territorial development wherein trout farming can have a critical role contributing to the socio-economic fabric of rural communities.

However, their political work on these issues has met with an ambiguous response. On the one hand, no French strategy for growth equivalent to the Scottish one (with growth targets and measures) was developed within the Ministry of Agriculture during this period. On the other hand, officials from within the Ministry of Ecology dispute that ecocentric visions of rivers are overtaking anthropocentric ones. On the contrary, they continue to argue that the application of the WFD has resulted from an extensive consultation of many stakeholders within the norms of the over-arching French environmental programme (known as Grenelle), and as such carries a societal legitimacy in which trout farming still has a role to play. They have especially stressed that funds have been put aside to help fish farmers make the necessary adaptation to meet any conditions imposed:

> I am not in favour of the death of fish farming in France, far from it. I am here to work transparently…. But I can't make legal texts say things they don't actually say.
>
> (Interview: Ministry of Ecology official, 2012)[23]

> It is not environmental norms which will provoke the death of fish farming. That is however what some people say. Of course, I understand

that French norms are sometimes stricter than those in other European countries, but we provide enormous support.

(Interview: Ministry of Ecology official, 2012)[24]

Strong tensions thus characterise the sustainability of access to fish farm sites in the case of Aquitaine trout, with some actors seeking to open this phenomenon up to political controversy and others seeking to close it down. In this climate, local prefectures and administrative judges scrutinising applications for sites have become increasingly cautious in their interpretations of the rules, often defaulting to principles of precaution in cases of uncertainty. This has led to a stalemate situation where producers no longer apply for (re)new(ed) authorisations, fearing either attack (e.g. by recreational fishers during the public inquiry phase) or failure.

Unpacking interdependencies

Political work on sustainability has especially included mobilisation around a politics at the frontier of different territorial visions, building alliances with other international organisations (e.g. IUCN) to develop a French aquaculture strategy, and the generation of new knowledge.

Territory: As was the case for Scottish salmon, Aquitaine trout frontier politics within interdependence and territory has emerged in conflicts between actors over how they have politically aligned visions of territorial and river basin futures and representations of the environment (biophysical processes and landscape aesthetics). A dominant vision of France and rivers' futures has been supported by a strong coalition of actors implementing the WFD (e.g. actors within the Ministry of Ecology, water agencies, ONEMA and river basin committees) as well as other stakeholders (e.g. recreational fishers, and national associations for the protection of water and rivers). This vision is a powerful one, rooted in the recent history of France's relationship to rivers and its past institutionalisation of river development projects and French national identity (Pritchard, 2004). The modification of rivers like the Rhône through intensive dam construction had been part of large development projects building state power through the 'nationalisation' of nature and described using metaphors of conquest (Pritchard, 2004). In the WFD, many actors saw the opportunity to reverse this process through a 'reconquest' (reconquête) of nature, removing dams and restoring rivers as part of a renewed French identity. This has been supported by a powerful coalition of actors legitimised through EU policy resources, and has contributed to the institutionalisation of quasi-ecocentric ecosystem approaches to water governance.

They call it reconquering nature ... Reconquest is a war term ... the reconquest of nature.

(Interview: French trout producer, 2012)[25]

Yet, whereas this vision has been clear concerning its representation of aquatic ecosystems, it has been more ambiguous over the role for trout farming in these rivers' futures:

> One can imagine that human activities take priority in some places – where there is a strong economic/social need – but for other rivers there is no reason not to respect these limits and reach good biological status – fauna and flora.
>
> (Interview: French public actor, 2012)[26]

Within this vision, trout producers perceive a prioritisation of nature as cultural benefits over provisioning ones (i.e. the reverse of the Scottish case). For example, they denounce:

> Environmentalist lobbyists who want to turn our lands into natural museums and economic deserts.
>
> (CIPA and FFA, 2007: 3)[27]

To keep alive arguments premised on more anthropocentric interpretations of ecosystem approaches, producers have engaged in extensive political work seeking to promote their visions of territorial futures and their representations of the environment working at a regional and local scale.

> When restructuring territory, you not only have to take the economic importance of industries into account, but also their relative environmental impact, their role in creating and maintaining employment in rural areas, as well as territorial autonomy in food security.
>
> (IUCN, 2011: 44)[28]

> Today the issue is how to perpetuate existing activities and to avoid breaking things which exist in the first place.
>
> (Interview: French trout producer, 2012)[29]

This political work has been advanced through multi-positioning strategies at EU, French, regional and local scales. As well as engaging in EU-wide consultations over European strategies for sustainable growth (and in the EU producers' association FEAP) and through CIPA at the French-wide scale, local representative associations, such as GDSAA have sought to carry their arguments through local and regional discussions in the river basin committees. As noted in Chapter 4, this has been challenging, not least because trout farming is a relatively small industry (GDSAA is 1 member out of 135 in the river basin committee).

In addition to this, local companies have worked on their insertion in local communities, including in some cases linking the fish farm to a recreational role, through providing opportunities for recreational fishing, or through river re-stocking practices.[30] A critical part of this work has been to invest in local commercial practices which I discuss in more detail in Chapter 6. Overall, work at the frontiers of territory has underlined the need to implement policies which stress territorial diversity, thereby moving away from 'black and white' state ministry positions (Germaine and Barraud, 2013).

Regulation: This political work managing interdependence and territory has been accompanied by one working at the interface of public/private interdependencies. First, one of the central challenges faced by trout producers has been that whereas a strong public policy towards water quality and ecological continuity in the form of the WFD has been institutionalised, there has been no equivalent public policy to grow trout farming aquaculture. Of course, trout farmers are in regular contact with officials from the fish farming division of the Ministry of Agriculture, who also act as an interface between them and the Ministry of Ecology.

Faced with a number of challenges influencing Ministerial policy through these channels, in 2010 trout farmers, working with officials in the fish farming division of the Ministry of Agriculture, mobilised specifically on this issue. This mobilisation included the politicisation of the absence of a strong Ministerial signal to grow the industry and the building of alliances with international e-NGOs, as well as French scientists (including both natural scientists and sociologists[31]), to set out the main elements of a strategy for its sustainable development.

This has been carried out on a number of fronts. For example, officials in the fish farming division of the Ministry of Agriculture had seen previous work conducted by the international NGO IUCN on Spanish aquaculture and contacted them to request that a similar exercise be carried out on French trout. This involved the holding of working groups bringing together industry actors and IUCN to discuss major issues facing the sustainability of trout farming and led to the writing of a detailed report (IUCN, 2011). In this report, sustainability was discussed in ecosystem services language, whereby a consistent proposal supported by the IUCN was to carry out ecosystem services' valuation as a possible regulatory response to inform public choices on trout farming in line with nature's societal benefits (IUCN, 2011: 54).

However, an important challenge remained bringing collective private actors' social meanings of ecology–society interdependencies into the heart of the sustainability institution. This was because, on the one hand, no-one from the Ministry of Ecology was present at the working groups or meetings preparing the IUCN report:

> In this initiative the Ministry of Ecology and Sustainable Development operated an empty chair policy.
>
> (Interview: French public actor, 2012)[32]

On the other hand, officials in the fish farming division of the Ministry of Agriculture found routine inter-Ministerial communication problematic. To redress this, a second initiative was put in place signing a charter between the two Ministries which set up a formal annual mechanism for inter-service communication:

> Faced with the challenge of making ourselves heard we had the idea of putting a charter in place.
>
> > (Interview: French public actor, 2012)[33]

These initiatives notwithstanding, political work over this interdependence has been a struggle. As summarised in the IUCN report representing trout farmers in their political work as 'warriors' (guerriers):

> The warriors are tired.
>
> > (IUCN, 2011; 16)[34]

Knowledge: Finally, as we saw in Chapter 4, dominant knowledge has been ecological knowledge setting indicators to implement the WFD (Bouleau and Pont, 2015). One important knowledge gap identified by actors has been the lack of sufficient technical support for trout farmers responding to the administration's demands (compared with agriculture, for example). According to the IUCN report (2011), technical advice is limited (one engineer at ITAVI)

Table 5.2 Summary of political work over Aquitaine trout access to farm sites

Aquitaine trout			
Social constructions of sustainability	Tensions between anthropocentric ecosystem approaches versus quasi-ecocentric ecosystem approaches. These tensions are opened up to controversy by producers and closed down to controversy by public officials in the Ministry of Ecology		
Frontier politics of polity interdependencies	Territory	Regulation	Knowledge
	French-wide approach to water governance dominates in line with renewed French identity; Political work by producers in alliance with other actors to politicise differences over how different actors have connected visions of territorial futures with representations of the environment	Absence of strong public signal to grow the industry politicised through public-private mobilisation calling for new regulatory responses within ecosystem services' valuation; Tensions over nature as providing cultural services at the expense of economic activities	Dominant regulative science; Technical knowledge gaps politicised but no destabilisation of hierarchies of knowledge

and other local officials in charge of the application of the law often have little knowledge about trout farming. One proposal emerging was to grant fish farming access to the services of Agricultural Chambers to increase technical and administrative support. Beyond this, trout farmers working through their collective associations have sought to communicate data on job creation and, as we saw in Chapter 4, have put in place measurement tools to provide objective data on environmental performance of their farms. However, even though this knowledge has been generated, it has not yet been taken up through political mobilisation to destabilise dominant knowledge forms and provide the sustainability institution with its social meaning.

Greece and governing access to fish farm sites

Overview of the changing sustainability narratives

The history of governing access to fish farm sites in Greece can be characterised as increasingly conflictual. Illegal planning and clashes over coastal futures are key elements in this story. Access to fish farms sites was governed in Greece for many years through institutional work led by central government Ministry officials and producers stabilising ecological modernisation approaches as they sought win–win outcomes in technical innovation and ecological benefits. Many licences were up for grabs in the early years and many companies took advantage of a favourable growth environment. However, the legitimacy of governing farm sites through public action depoliticising ecology–economy interdependencies has increasingly been threatened through the action of local communities, leading to the annulment of some producer licences by the Greek State Council.[35] The consequence has been the halting of growth and the forcing of sales of fish. This has led eventually to a re-institutionalisation of sustainability using policy tools developed from within ecosystem approaches to spatial planning.

The development of the industry in Greek coastal communities initially experienced very fast rates of growth. For example, between 1990 and 1999, the average annual increase in production was 70 per cent (European Commission 2009b: 64). More recently, this form of coastal development has increasingly been challenged by local communities. At first glance, it appears as if local communities' lack of support for fish farming has been connected to issues of trust – both of producers complying with rules and of the state monitoring farm sites (see Chapter 4). These reasons notwithstanding, a deeper issue of the societal benefits of nature underlies communities' criticisms of fish farming. Rather than aquacultural development in the lagoons, many Greeks prefer tourism and consider the two to be incompatible:

There are a lot people raising voices against aquaculture. There is a lot of public pressure, even from fishermen, from people who own hotels, from local people that do not work in aquaculture.

(Interview: Greek scientist, 2012)

Indeed, many of the challengers have been keen to develop tourism and have considered that the presence of fish farms would be problematic for this purpose, diminishing the value of their land. However, this has not been a straightforward case of land use conflicts either, because initially fish farms were not developed in touristic areas. Nonetheless, as tourism began to spread, the presence of farms became problematic. Interested parties began to complain to the planning department in the Greek government and raise their concerns in the State Council. In response, the State Council began annulling producer licences.

Critically, this annulment of licences did not result from non-compliance by producers with environmental rules. Rather, it resulted from the fact that, over time, successive Greek governments had failed to develop enabling framework legislation for planning. In fact, they had allocated licences to farms on a case by case basis and in the absence of a national spatial planning framework. According to the law, many of the texts on which licences were awarded were not legally effective as they depended on this spatial planning document (Conan and Prieur, 2000). Annulment of licences was therefore an act of judicial review against the Greek government and not an act of environmental control over producers:[36]

> The most basic problem is that we don't have any kind of spatial planning laws – there was no national zoning plan in Greece. The constitution requires us to have this. It has been in the constitution for thirty years – not just for aquaculture, for everything.
> (Interview: Greek seabass/seabream producer, 2012)

Greek officials working with producers, research institutes and in consultation with the public have sought to find solutions to these attacks which have destabilised the win–win ecology–economy narrative. These solutions have been found in the development of what officials in the Ministry for the Environment, Spatial Planning and Public Works have described as a new 'integrated marine spatial plan' (Karka, 2011).[37] Consequently, following the reframing of the debate around ecosystem benefits coupled with the implementation of an integrated plan, it appears that Greek sustainability norms governing access to farm sites are currently being re-institutionalised around anthropocentric ecosystem approaches.[38] Indeed, institutional work over the contents of the plan has created new conflicts within a new framing of the sustainability institution concerning ecosystem approaches to services and who benefits from nature. Should seas and coasts benefit tourism and local communities or fish farmers? Can a synergy be found between interacting ecosystem services?

Unpacking interdependencies

To further grasp the transformation of the sustainability institution and its current political conflicts, we must consider actors' political work over the frontier politics of interdependence and territory, regulation and knowledge.

Territory: As was the case for Scottish salmon and Aquitaine trout, politics at the frontiers of territory have centred upon conflicts between actors over how visions of coastal futures and representations of the environment (biophysical and landscape aesthetics) have been aligned. On the one hand, the Greek government and producers have considered that seabass and seabream farming has an important role to play in Greek coastal futures. On the other hand, representations of some Greek coastal communities do not share this belief. In both cases, territory has become an important resource in these conflicts.

First, a central element of the Greek government's response to attacks against the sustainability institution has been to mobilise EU institutional resources in the form of a policy tool on integrated marine planning. The decision to do so stemmed partly from initiatives within the Ministry for the Environment, Spatial Planning and Public Works but also importantly from direct intervention by the European Commission. Indeed, the European Commission had become frustrated through having invested in the development of the sector through EU funding, only to find that licences were subsequently being annulled. It placed a moratorium on releasing funds under the framework programme until the problem was resolved. In this case, therefore, not only was EU territory 'brought in' to resolve a local problem (Pasquier, 2005), but also EU public actors directly intervened in Greek domestic politics. The introduction of the plan also brought into the sustainability institution EU norms on participatory democracy. This has been described as part and parcel of the procedures developing integrated marine spatial planning (Stead *et al.*, 2002).

Territory has also been present in arguments and strategies of those attacking seabass/seabream farming. Symbolically, this has created situations where coastal villages have hung up black flags in the village because all the locals are against the fish farm. A seabass/seabream farm site manager described the problems they had with their neighbours:

> The artists living on the hill are against us, the mayor and the prefect are against us, whereas in the past they had been in favour of us and had invited our business which created local jobs. Now the mayor wants to replace us. We've been attacked in court, but we expect that the plan will now protect us.
>
> (Interview: Greek farm site manager, 2012)

> It is due to the perception of land use conflict. We have 13,000 kilometres of coastline in Greece. All of the fish farms in Greece is seven square kilometres put together. All of them. We take up a tiny little part of the coastline. This is why I say perceived. Because we are not zoned anywhere which are zoned for tourists. This was clearly the places for tourism and there is no farming there. There is the perceived sense that some day somebody will come and make a huge resort here … a hope of 'if I have a little piece of farmland and one day someone will come and

make use of my land and if there is a fish farm then the value of my land goes down'.

(Interview: Greek seabass/seabream producer, 2012)

This has led to clashes between producers and land users, local authorities and tourism interests (Papoutsoglou, 2000: 171). These conflicts over interactions of nature and local territorial futures at the local scale play out over alternatives. In Greece, tourism represents for many people a viable alternative future in the local economy, and this has at times been coupled with protectionist local attitudes.

Regulation: Political work around the interface of territorial interdependencies has been cross-cut by a political work altering the frontiers of state/civil society interdependencies. Already in 2000, Conan and Prieur (2000) observed that the Greek fish farm location authorisation process had resulted in the creation of a great many legal texts, but that these were not legally effective as they depended on a spatial planning document which had not been developed. In the absence of state action to regulate the planning through the development of a framework planning regulation, private individuals have taken action against the state (resulting in the State Council annulling the licences).

> The problem of acceptance from local societies was a very big problem when aquaculture started. The local people didn't want a farm there, the local fishermen didn't want a farm there, so local people started going to the police to ask them to go and check. Go and see. He has a licence. What is his fish? What is he feeding his fish. And this continues nowadays. A system of control mainly from the local people who couldn't understand the perspective of the industry. And they involved the administration over all decisions.
>
> (Interview: Greek seabass/seabream producer, 2012)

> Here in Greece we are making always the same mistake. We are starting always from the end. We are developing also the sector and the industry with 320 licences available in the whole of Greece and then we started speaking about how these farms are going to work together with tourism and other things. So we start at the end to make a planning.
>
> (Interview: Greek seabass/seabream producer, 2012)

> After we put the farm in the sea and then we ask should it be there or should we move it there? This gives a lot of pressure for the farms in the local communities where they operate, it creates a lot of conflicts especially for the people working there, but it also makes a cost because you have to have a lot of lawyers and councils and consultants.
>
> (Interview: Greek seabass/seabream producer, 2012)

As I have argued, the response to this problem has been the development of a Greek Special Framework for Aquaculture in 2011.[39] The plan designated different zoned coastal areas organised into several categories, namely: (i) Areas suitable for the development of aquaculture; (ii) Allocated zones organised and managed by a special authority, similar to industrial estates; (iii) Informal zones in places with existing concentrations; (iv) Individual location: (a) within suitable zones and (b) outside the provided zones (Karka, 2011: 7). The objectives of the plan were simultaneously to zone areas for aquaculture production and to manage legislative anarchy (for example through creating legal certainty and reducing the potential for non-compliance, blackmail and corruption). There was also a high expectation by central Ministries that the new plan would diminish controversies between local authorities and communities.

However, the implementation of the plan has engendered new conflicts, this time between producers and the planning ministry. Two issues have emerged. First, there is ambiguity over whether the plan aims at growing the industry, or whether it is a plan aimed at categorising already existing activities. According to the Ministry, the plan is about growth and development in those areas it has designated, but not all producers see it this way. They have argued instead that the plan is not based on a vision for industry. In response, government officials have retorted that this is because companies are requesting very large sites from 250 tonnes up to 1800 tonnes and do not accept the limits being imposed on them. Second, there has been a fierce debate about the 'integrated' dimension of the plan and whether it is in fact an integrated plan or not:

> It is not to integrate activities, it is to isolate activities.
> (Interview: Greek seabass/seabream producer, 2012)

According to producers, this issue has emerged because of the way in which the plan has carved out special zones for aquaculture, whilst simultaneously excluding other activities from these areas. Whereas this form of zoning can work for land management, it has been criticised when applied to the sea, which is a multi-dimensional space. There have been exchanges over the integration of fish farming with two other activities: fisheries and tourism. On fisheries, producers have argued that the zoning laws exclude fisheries around the cages, but they do not understand why and have expressed concerns that this gives the impression that the fish farms are bad for fisheries:

> The other argument we had was over fisheries. I said we have to integrate with the local fishing communities, we are not competitive industries, we are complementary. Their industry is dwindling and our industry is increasing and we have to find ways to work together and very often where you have a fish farm where the feed goes in you have an increase in fish stocks. So this could be good for both of us, so you

mustn't exclude them. For example, the zoning laws we have excludes fishing activities around the cages.

(Interview: Greek seabass/seabream producer, 2012)

This puts a label on the industry, which sometimes is bad. Why can't they fish around [the cages]? Why not have fishing here and a farm in the next bay? It is not allowed, which means that fish farming is bad for tourism, for the bathers, which it is not. But if by law we exclude, then it puts a label on *my* industry that it is bad.

(Interview: Greek seabass/seabream producer, 2012)

But why can't they fish next to the cages? There is no problem with that – there is more fish around the cages. Here we let the fishermen fish around the cages. In the summer there is a boat and we let them visit to see how it is done. These activities should be integrated.

(Interview: Greek seabass/seabream producer, 2012)

There have also been conflicts over the integration of seabass/seabream farming and tourism. Of course, behind the idea of the plan was the objective to clearly delineate areas for fish farming as distinct from areas for tourism. But this, it is argued, has now precluded other models of coastal development from being put in place:

It creates conflicts. One of the arguments I had with the Ministry of the Environment was that this separation of tourism and aquaculture – if the tourism owner wants to have an aquaculture company, it should be allowed. So I am thinking of the example of a small hotel which has organic food, organic oil, why can it not have two cages of organic fish? Why can this not be a model of tourism and development if the hotel owner wants it.

(Interview: Greek seabass/seabream producer, 2012)

In response to private judicial action, therefore, public regulation has been reinforced through the development of the new framework plan for aquaculture. However, the development of this new plan has revealed a shift in the governing approach whereby the plan has been developed through a form of participatory democracy, acknowledging changing relations between state and civil society. This appears to be re-institutionalising the sustainability institution around anthropocentric ecosystem approaches, thereby opening debates on policy integration and how to give this meaningful effect.

Knowledge: Finally, at the heart of this changing sustainability institution I observed political work over interdependence and knowledge. Political work by officials within the central Greek government's Ministry for the Environment, Spatial Planning and Public Works to draw up the plan in 2011 has

Table 5.3 Summary of political work over Greek seabass and seabream access to farm sites

Greek seabass and seabream			
Social constructions of sustainability	From de-politicised ecology–economy interdependencies to politicised anthropocentric ecosystem approaches. Tensions over which economic activities should benefit nature and how to govern interactions		
Frontier politics of polity interdependencies	Territory	Regulation	Knowledge
	Territory as a critical resource in conflicts between actors over how visions of coastal futures and representations of the environment have been aligned	Initial attacks on the sustainability institution through protests have been depoliticised through alteration of the frontiers between state and civil society; New form of participatory democracy institutionalised	Scientification of controversies

been accompanied by the generation of new knowledge. In particular, to legitimise the plan, different types of knowledge have been gathered by officials managing tensions within the sustainability institution.

As was the case for the politics of access to farm sites in Scottish salmon, this political work can be described as a 'scientification of controversies' (Weingart, 1999). First, two different studies informed the plan: a study financed by the Federation of Greek Maricultures (FGM) and a study commissioned by the Minister. Second, a public consultation was organised which received 768 responses, thus co-opting local communities' views regarding the re-institutionalisation of the sustainability institution towards anthropocentric ecosystem approaches. These different knowledge types have since been evoked by Ministry officials legitimising how the plan seeks to find a balance between coastal nature as a societal benefit for seabass and seabream production versus coastal nature as a societal benefit for tourism.[40]

Conclusions

In this chapter, I have examined how through their political work, actors in Scottish salmon, Aquitaine trout and Greek seabass and seabream have engaged in the governing of access to farm sites. As set out at the beginning of the chapter, access to farm sites could potentially be problematised in a number of different ways connected to the overall insertion of fish farming aquaculture in local territorial development. These included constructing the societal worth of fish farming aquaculture as a response to central challenges posed in the Anthropocene; applying integrated policy tools for local community discussions on the interaction of this economic activity with other economic activities impacting on the same environment; and moving off-shore. Applying the framework we can see that there is nothing inevitable about the way actors have framed the issue, nor about how policy tools actually work in practice. In the case studies quite different social constructions of access to farm sites inform quite different policy instruments implemented. This is to reveal a comparative politics over the governing of environmental landscape aesthetics and nature as a societal benefit.

In Scottish salmon, planning and siting decisions were initially governed from within a sustainability institution espousing depoliticised ecology–economy interdependencies. The contents of the sustainability institution have since changed, however, whereby access to farm sites is currently regulated through an anthropocentric ecosystem approach. Change was facilitated through the separation of a Scottish approach post-devolution which permitted the emergence of a new coalition of actors shifting planning from private to public control. Political work by producers and the SG versus regulatory agencies and NGOs has centred upon farm siting, taking into consideration not only interactions with wild fisheries but also landscape aesthetics. More recently, tensions have emerged between these actors over whether planning regulation is continuing to support nature as providing multiple benefits to Scottish society, or whether it is being applied in a manner which contributes instead to it predominantly supporting nature providing (economic) benefits to dominant salmon companies.

By contrast, in Aquitaine trout, access to farm sites has been governed from within a sustainability institution dominated by actor tensions over anthropocentric ecosystem approaches versus quasi-ecocentric ecosystem approaches. Whereas producers (at times working in alliance with officials from the fish farming division of the Ministry of Agriculture and/or international NGOs) have sought to open up these tensions to controversy, officials within the Ministry of Ecology have consistently sought to close them down.

The politics of sustainability has therefore been strongly cross-cut by political work led by producers (in alliance with other actors) to politicise differences over how actors have connected visions of territorial futures with representations of the environment; as well as public/private mobilisation calling for new regulatory responses within ecosystem services' valuation and the politicisation of technical knowledge gaps. However, for now, as was the case over fish farm/environment interactions, ambiguities persist regarding the authority of 'sector' actors over their own destiny. The question remains as to whether the social construction of rivers for cultural benefits will ultimately come at the expense of trout farming.

In Greek seabass and seabream, the initial governing of access to farm sites took place from within a sustainability institution depoliticising ecology–economy interdependencies. However, this institution was increasingly attacked through civil action disrupting it. Officials within the Greek government have sought to re-institutionalise sustainability, this time giving it social meaning through anthropocentric ecosystem approaches which problematise ecology–society interdependencies. This has led to the use of integrated planning tools. In this re-institutionalisation, territory has been a critical resource in conflicts over visions of sustainable coastal futures, although the more recent separation of state autonomy from civil society has enabled new forms of participatory democracy to be established. However, as was the case for both Scotland and Aquitaine, new tensions have emerged over 'which' economic activities nature should benefit and the limits of spatial planning tools to govern ecosystem services' conflicts.

These findings permit us to draw some more general conclusions comparing European aquaculture policy on access to fish farm sites. To begin with we can observe that the objectives of these policies differ between territories and sectors. Whereas we can describe Scottish policy as one of growth, the French policy applied in Aquitaine is rather one of control, whilst the policy in Greece is more one of containment in specified zones of production. In each case, these policy objectives are contested and subject to compensation.

In Scotland, the main policy instruments for compensating growth have been set within the planning system and extended beyond questions of environmental and biological impacts to include visual and aesthetic impact assessments. Yet these compensatory measures have not been sufficient in all communities and amongst all stakeholders in mitigating tensions over choices for coastal futures. A central challenge has been that discussions over planning have taken place in two different participatory governing spaces which do not always join up: within the working groups on planning, on the one hand, and within local authority debates on case by case planning applications, on the other.

In Aquitaine, the near hijacking of licence authorisation processes by interested e-NGOs which have attacked farm applications on a case by case basis has resulted in no new applications for new sites being sought over decades. In this climate, successive French governments have made ambiguous statements about whether fish farming aquaculture should be growing or not (in Aquitaine or in France in general), and no public democratic space has been opened up specifically dedicated to debating the future of trout farming. In its absence, producers working collectively and in liaison with international NGOs have created their own 'spaces' for debate and for constructing the lack of growth (and further decline) as both an industry-wide and public policy problem which needs to be addressed.

In Greece, early policy mechanisms of licensing enabling growth reached their limits in meeting policy objectives due to their being issued without accompanying state regulation to provide them with legality. Following private and judicial activism challenging their legality, where debates on aquaculture's future were decided upon by tribunals, new spatial planning tools have now been applied to plan industry and coastal development. These have made it possible to bring the problem of growth into a regulatory public space of action (as distinct from a space of civil protest), subject to norms of participatory democracy. The plan has not, however, resolved tensions, which continue over the place of fish farming in Greek coastal futures.

Notes

1 See, for example, the Tanguy report 2008 on the future of aquaculture in France (Tanguy, 2008).
2 See: www.gaaia.org/salmon-pharming.
3 For a discussion of the genesis of ICZM, see Ségalini, 2013; and on its application, see McKenna *et al.*, 2008.
4 This also includes making changes in production away from monocultures to integrated aquaculture production systems (Costa-Pierce, 2010).
5 Including the institutional work on products, whereby fish farmers would be encouraged to produce a diversity of products, including for local markets (Costa-Pierce, 2010: 100–101); see also Chapter 6.
6 See Chapter 4 for a discussion.
7 The procedure in Shetland was slightly different due to differences in rules over land and sea ownership. It was described during interview as follows:

> Our local authority sitting down with industry wanted to create a policy for the development of the farms; so very quickly we created a local process by which farms could be applied for – a development process – to get a works' licence for your activity. So we created all these things very much to suit the islands' requirements. It is a bit different from the rest of Scotland. The rest of Scotland went down a slightly different route where they had to apply to the Crown Estate to get a lease to operate. We still had to do that, but we had gone beyond that to actually saying 'we want to control that and we want to be part of the control process'.
>
> (Interview: Salmon producers' association, 2012)

8 Except for radical e-NGOs, which have been against salmon production from the start.
9 Following the conclusions of the Montgomery Committee of Inquiry (Macartney, 1985) into the functions and powers of the Islands Councils and the subsequent recommendations of the Scottish Development Department. Consultees included existing leaseholders; tenants; landowners; fishers; navigation; other departments and agencies; local authorities; conservation societies; general public; and the Crown Estate Commissioners.
10 In these debates, nature was constructed as having multiple benefits within notions of ecosystem services.
11 Through the development of new environmental 'visual impact assessments' – see below.
12 Currently, this plan is to 'increase [salmon production] sustainably' to 210,000 tonnes by 2020 (DEFRA, 2015). An ad hoc working group of industry and public/private organisations has more recently released a renewed strategic plan for growth to 2030 (mentioning figures of 300,000–400,000 tonnes per annum) (Scotland Food and Drink, 2016).
13 Specifically, between marine waters as provisional services for aquaculture versus marine waters as cultural services for tourism.
14 Scottish Parliament.
15 Anecdotes in interviews confirm these arguments that fish farming sites provide high levels of local employment in very rural areas with tiny communities, e.g. in Scourie (NE Scotland), 60 jobs; in Stornaway, 100 jobs.
16 For example, whilst not saying a categorical 'no' to salmon farming, wild salmon fisheries' boards have been keen to ensure that its expansion is site-sensitive to reduce interactions with wild fish.
17 For many actors these two processes went hand in hand.
18 Setting up a Special Task Force which sat 'above' the working groups.

19 This marked the start of the setting up of the cooperative Aqualande, whose very origin in 1981 was concerned with social as well as economic objectives in local territorial development (interviews).

20 Average French decline during this period was −20 per cent (Agreste, 2011).

21 Original French: 'On est sur une vision particulière de l'action administrative et environnementaliste ... On a fait une réglementation environnementale qui n'est pas une réglementation durable. Aujourd'hui si on applique à la lettre le règlement on ferme la moitié des piscicultures en France.'

22 Original French: 'La question c'est est-ce qu'on met l'homme dans l'environnement ou est-ce qu'on l'extirpe de l'environnement? ... [On] a tendance à exclure l'homme de l'environnement'.

23 Original French: 'Moi je ne suis pas là pour la mort de la pisciculture en France, bien loin de cette idée. Je suis là pour travailler en transparence.... Mais je ne peux pas faire dire aux textes les choses qu'ils ne disent pas.'

24 Original French: 'C'est pas les normes environnementales qui vont provoquer la mort de la pisciculture. C'est quand même ce qui est dit par certains. Après je veux bien entendre que les normes françaises sont parfois plus sévères que les normes d'autre pays européens, mais on aide vachement.'

25 Original French: 'On appelle ça la reconquête du milieu. La reconquête c'est un terme de guerre ... la reconquête du milieu'.

26 Original French: 'On peut estimer que les activités humaines priment dans certains endroits – ou il y a un besoin fort économique/social – mais pour les autres rivières il n'y pas de raison de ne pas respecter ces paramètres et atteindre les seuils de bon état biologique – faune et flore.'

27 Original French: 'Les lobbies environnementalistes qui veulent faire de nos pays des musées naturalistes et des déserts économiques.'

28 Original French: 'Il s'agit notamment de prendre en compte dans l'aménagement du territoire non pas seulement l'importance économique des filières, mais aussi leur l'impact relatif sur l'environnement, leur rôle dans le maintien et la création d'emplois en milieu rural, ainsi que les enjeux en matière d'autonomie alimentaire des territoires.'

29 Original French: 'Aujourd'hui l'enjeu est de les activités existantes, d'éviter de casser les choses qui existent, dans un premier temps.'

30 River re-stocking practices are also quite controversial, crystallising the issue being described here. This includes debate on whether or not the introduction of certain species of non-native rainbow trout (which have been sterilised, but which can be important predators) into the rivers for recreational fishing purposes is in keeping or not with policies of naturalisation of rivers.

31 Known as the GIS-pisciculture initiatives (le GIS Piscicultures Demain: GIS PsD).

32 Original French: 'Dans cette démarche le ministère de l'écologie et du développement durable a fait la politique de la chaise vide.'

33 Original French: 'Face à la difficulté de se faire entendre, l'idée est venue de mettre en place la charte.'

34 Original French: 'Les guerriers sont fatigués.'

35 The highest administrative court in Greece.

36 Of course, this implied that all licences in Greece were technically illegal in the sense that none of them had been granted within the necessary legislative framework.

37 According to Karka, the Aquaculture Framework is one element in a larger national programme on spatial planning developed by the former Ministry for the Environment, Spatial Planning and Public Works. Other plans include the General (National) Framework on Spatial Planning and Sustainable Development and sectoral spatial plans on Renewable Energy Resources, Industry and Tourism (2011: 3).

38 In his country review for the Food and Agriculture Organisation on ecosystem approaches to aquaculture, Costa-Pierce made a recommendation for coastal zoning of aquaculture, moving towards an ecosystem approach in Greece (2008: 94).
39 In fact this replaces the 2003 Special Framework for coastal areas which had been developed but never endorsed.
40 In the language of ecosystem services, this refers to tensions between 'provisioning' versus 'cultural' aquatic ecosystem services.

References

Agreste, 2011. 'La salmoniculture, un secteur en recul', *Les Dossiers N° 11*, April 2011.

Bailey, C. 2008. 'Human dimensions of an ecosystem approach to aquaculture'. In Soto, D., Aguilar-Manjarrez, J. and Hishamunda, N. (eds) *Building an Ecosystem Approach to Aquaculture*, FAO/Universitat de les Illes Balears Experts Workshop. 7–11 May 2007, Palma de Mallorca, Spain. FAO Fisheries and Aquaculture Proceedings No. 14, Rome: Food and Agriculture Organisation (FAO), pp. 37–46.

Bermúdez, J. 2008. 'Legal implications of an ecosystem approach to aquaculture'. In Soto, D., Aguilar-Manjarrez, J. and Hishamunda, N. (eds) *Building an Ecosystem Approach to Aquaculture*, FAO/Universitat de les Illes Balears Experts Workshop. 7–11 May 2007, Palma de Mallorca, Spain. FAO Fisheries and Aquaculture Proceedings No. 14, Rome: Food and Agriculture Organisation (FAO), pp. 67–78.

Bouleau, G., Pont, D., 2015. 'Did you say reference conditions? Ecological and socio-economic perspectives on the European Water Framework Directive', *Environmental Science & Policy*, 47: 32–41.

Chang, B., Coombs, K., Page, F. 2014. 'The development of the salmon aquaculture industry in southwestern New Brunswick, Bay of Fundy, including steps toward integrated coastal zone management', *Aquaculture Economics & Management*, 18(1): 1–27.

CIPA (Comité Interprofessionnel des Produits de l'Aquaculture) and FFA (Fédération Française de l'Aquaculture). 2007. 'Contribution de la filière aquacole française à la consultation lancée par la Commission européenne sur les perspectives de développement de l'aquaculture communautaire', written contribution to the European Commission's external consultation on the renewal of the European Strategy on the Sustainable Development of Aquaculture, July, 2007.

Conan, N., Prieur, L. 2000. 'Les piscicultures marines et l'environnement en Grèce'. In Petit, J. (ed.) *Environnement et aquaculture, Tome II Aspects juridiques et réglementaires*, Paris: INRA, pp. 257–266.

Costa-Pierce, B. 2008. 'An ecosystem approach to marine aquaculture: a global review'. In Soto, D., Aguilar-Manjarrez, J. and Hishamunda, N. (eds) *Building an Ecosystem Approach to Aquaculture*, FAO/Universitat de les Illes Balears Experts Workshop. 7–11 May 2007, Palma de Mallorca, Spain. FAO Fisheries and Aquaculture Proceedings No. 14, Rome: Food and Agriculture Organisation (FAO), pp. 81–115.

Costa-Pierce, B. 2010. 'Sustainable ecological aquaculture systems: the need for a new social contract for aquaculture development', *Marine Technology Society Journal*, 44(3): 88–112.

Coull, J. 1988. 'Fish farming in the Highlands and Islands: boom industry of the 1980's', *Scottish Geographical Magazine*, 104: 4–13.

Culver, K., Castle, D. (eds). 2008. *Aquaculture, Innovation and Social Transformation*, Springer International Publishing.

DEFRA (Department for Environment, Food and Rural Affairs). 2015. *United Kingdom Multiannual National Plan for the Development of Sustainable Aquaculture.* Available at: www.gov.uk/government/publications.

Ernst & Young. 2008. *Etude des performances économiques et de la compétitivité de l'aquaculture de l'Union européenne,* Report for DG MARE, December.

European Commission of the European Communities. 2009a. *Building a Sustainable Future for Aquaculture,* Communication, Com(2009)162, 8th April, 2009, Brussels.

European Commission of the European Communities. 2009b. *Impact Assessment in Respect of Com(2009)162,* SEC(2009)453, Brussels.

Ferreira, J., Saurel, C., Lencart e Silva, J., Nunes, J., Vazquez, F. 2014. 'Modelling of interactions between inshore and offshore aquaculture', *Aquaculture,* 426/427: 154–164.

Germaine, M-A., Barraud, R. 2013. 'Les rivières de l'ouest de la France sont-elles seulement des infrastructures naturelles? Les modèles de gestion à l'épreuve de la directive-cadre sur l'eau', *Natures Sciences Sociétés,* 4(21): 373–384.

Grant, J. 2010. 'Coastal communities, participatory research, and far-field effects of aquaculture', *Aquaculture Environment Interactions,* 1: 85–93.

Hall, C., Scott, D., Gössling, S. 2013. 'The primacy of climate change for sustainable international tourism', *Sustainable Development,* 21: 112–121.

Holmer, M. 2010. 'Environmental issues of fish farming in offshore waters: perspectives, concerns and research needs', *Aquaculture Environment Interactions,* 1: 57–70.

IUCN. 2011. *Guide pour le développement durable de l'aquaculture: Réflexions et recommandations pour la pisciculture de truites.* Gland, Switzerland and Paris, France: IUCN.

Karka, L. 2011. 'Spatial Planning for Aquaculture: A Special National Framework for Resolving Local Conflicts', 51st Congress of the European Regional Science Association: Special Session Territorial Governance, Rural Areas and Local Agro Food Systems.

Lloyd, M., Livingstone, L. 1991. 'Marine fish farming in Scotland: proprietorial behaviour and the public interest', *Journal of Rural Studies,* 7(3): 253–263.

Macartney, A. 1985. *Summary of Findings of Montgomery Committee of Inquiry into Functions and Powers of Island Councils.* Available at: www.scottishgovernmentyearbooks. ed.ac.uk/record/22949?highlight=★:★.

Maréchal, G., Spanu, A. 2010. 'Les circuits courts favorisent-ils l'adoption de pratiques agricoles plus respectueuses de l'environnement', *Le courrier de l'environnement de l'INRA,* 59: 33–45.

McKenna, J., Cooper, A., O'Hagan, A.M. 2008. 'Managing by principles: a critical analysis of the European principles of Integrated Coastal Zone Management (ICZM)', *Marine Policy,* 32(6): 941–955.

Muir, J., Brugere, C., Young, J. Stewart, A. 1999. 'The solution to pollution? The value and limitations of environmental economics in guiding aquaculture development', *Aquaculture Economics & Management,* 3(1): 43–57.

Nunes, J.P., Ferreira, J.G., Bricker, S.B., O'Loan, B., Dabrowski, T., Dallaghan, B., Hawkins, A.J.S., O'Connor, B., O'Carroll, T. 2011. 'Towards an ecosystem approach to aquaculture: assessment of sustainable shellfish cultivation at different scales of space, time and complexity', *Aquaculture,* 315: 369–383.

Papoutsoglou, S. 2000. 'Monitoring and regulation of marine aquaculture in Greece: licensing regulatory control and monitoring guidelines and procedures', *Journal of Applied Ichthyology,* 16: 167–171.

Pasquier, R. 2005. '"Cognitive Europeanization" and the territorial effects of multilevel policy transfer: local development in French and Spanish regions', *Regional & Federal Studies,* 15(3): 295–310.

Peel, D., Lloyd, M. 2008. 'Governance and planning policy in the marine environment: regulating aquaculture in Scotland', *The Geographical Journal*, 174(4): 361–373.

Pillay, T. 1997. 'Economic and social dimensions of aquaculture management', *Aquaculture Economics & Management*, 1(1/2): 3–11.

Pritchard, S. 2004. 'Reconstructing the Rhône: the cultural politics of nature and nation in contemporary France, 1945–1997', *French Historical Studies*, 27(4): 765–799.

Rana, K. 2007. *Regional Review on Aquaculture Development 6: Western-European Region-2005*, FAO Fisheries Circular No. 1017/6, FIMA/C1017/6, Rome: Food and Agriculture Organisation (FAO).

Read, P., Fernandes, T., Miller, K. 2001. 'The derivation of scientific guidelines for best environmental practice for the monitoring and regulation of marine aquaculture in Europe', *Journal of Applied Ichthyology*, 17: 146–152.

Scotland Food and Drink (SFD). 2016. *Aquaculture Growth to 2030: A Strategic Plan for Farming Scotland's Seas*, Edinburgh: SFD.

Scottish Government. 2009. *A Fresh Start: The Renewed Strategic Framework for Scottish Aquaculture*. Edinburgh: Scottish Government.

Scottish Natural Heritage (SNH). 2002. *SNH's Vision of Sustainable Marine Aquaculture in Scotland: Consultation Response on a Strategic Framework for Aquaculture in Scotland*, Edinburgh: SNH.

Ségalini, C. 2013. 'Éléments de compréhension du processus de politisation du discours sur la gestion intégrée des zones côtières', *Développement durable et territoires* [online], 2(3): 1–10.

Shainee, M., Ellingsen, H., Leira, B., Fredheim, A. 2013. 'Design theory in offshore fish cage designing', *Aquaculture*, 392/395: 134–141.

Soto, D., Aguilar-Manjarrez, J., Brugère, C., Angel, D., Bailey, C., Black, K., Edwards, P., Costa-Pierce, B., Chopin, T., Deudero, S., Freeman, S., Hambrey, J., Hishamunda, N., Knowler, D., Silvert, W., Marba, N., Mathe, S., Norambuena, R., Simard, F., Tett, P., Troell, M., Wainberg, A. 2008. 'Applying an ecosystem-based approach to aquaculture: principles, scales, and some management measures'. In Soto, D., Aguilar-Manjarrez, J. and Hishamunda, N. (eds) *Building an Ecosystem Approach to Aquaculture*, FAO/Universitat de les Illes Balears Experts Workshop. 7–11 May 2007, Palma de Mallorca, Spain. FAO Fisheries and Aquaculture Proceedings No. 14, Rome: Food and Agriculture Organisation (FAO), pp. 15–35.

Stead, S., Burnell, G., Goulletquer, P. 2002. 'Aquaculture and its role in Integrated Coastal Zone Management', *Aquaculture International*, 10: 447–468.

Tanguy, H. 2008. Rapport final de la mission sur le développement de l'aquaculture. Ministère de l'Agriculture et de la Pêche. Ministère de l'Ecologie, de l'Energie, du Développement Durable et de l'Aménagement du Territoire. Available at: http://archives.agriculture.gouv.fr/publications/rapports/mission-sur/downloadFile/FichierAttache_1_f0/Rapport_H_Tanguy.pdf.

Weingart, P. 1999. 'Scientific expertise and political accountability: paradoxes of science in politics', *Science and Public Policy*, 26(3): 151–161.

Whitmarsh, D., Palmieri, M. 2009. 'Social acceptability of marine aquaculture: the use of survey-based methods for eliciting public and stakeholder preferences', *Marine Policy*, 33: 452–457.

Young, J., Brugere, C., Muir, J. 1999. 'Green grow the fishes-oh? Environmental attributes in marketing aquaculture products', *Aquaculture Economics & Management*, 3(1): 7–17.

6 Sustainability as a food governance problem

Product quality and shadow ecologies

Introduction

In this chapter, I examine the institutionalisation of sustainability as a food governance problem for Scottish salmon, Aquitaine trout and Greek seabream and seabass. In so doing, I connect issues associated with the growing of fish to those associated with its selling as a food product.[1] Of course, aquaculture is a food industry.[2] Yet the question can be raised whether, and how, sustainability is posed as a food governance problem. Whereas Chapter 4 covered sustainability issues cast by actors as problems from the farm scale and up to the water body/watershed scale, the following Chapter 5 covered sustainability issues cast by actors as problems from the water body/watershed scale and up to the community, regional and state-wide scales. What follows addresses sustainability issues potentially cast by actors as extending from the local market up to the global market scale.

The scalar parameters of this issue are global because of the trajectory of the 'tale' about aquaculture's 'natural' sustainable role: namely, because aquaculture compensates for collapsing fish stocks, farmed fish products can automatically be considered sustainable (Costa-Pierce, 2010: 88). As we have seen, although this story was challenged in the 1990s through increased awareness of the environmental impacts of farming, management measures implemented in response were sufficient to keep it alive (Young et al., 1999).

However, this story was to completely unravel on a global scale from 2000 onwards following the publication of a critical article by Naylor et al. in Nature (2000) which contested the very argument that aquaculture was compensating for collapsing sea fisheries. Naylor et al. set out what they termed a paradox in fish farming's relationship to fisheries (2000). Whereas aquaculture had hitherto been regarded as a solution for collapsing fish stocks, Naylor et al. argued that aquaculture was simultaneously *contributing* to collapsing fish stocks (2000: 1017). This was due to the contents of farmed fish diets. Farmed carnivorous fish (e.g. salmon, trout, seabass and seabream) are fed feeds which are made up of different ingredients: fish meal (FM); fish oil (FO); cereal sources; vegetable proteins, e.g. soya; vitamins; minerals; and feed additives.

At the time of the Naylor *et al.* publication, farmed fish were being fed on diets which contained high levels of FM and FO sourced from feed capture fisheries (for example, small pelagic fish like anchoveta from Peru, Chilean jack mackerel and NE Atlantic herring). Naylor *et al.* (2000) argued that the growth of aquaculture was putting direct pressure on the sustainability of these (and other feed fisheries) stocks: these feed fisheries were either at risk of collapse or had been fished to their limits. In other words, rather than benignly compensating for the loss of capture fisheries, fish farming aquaculture's dependence on capture fisheries for feeds was directly contributing to the unsustainability of such fisheries.

Presenting aquaculture's relationship to fisheries in this light resulted in the issue of farmed fish as sustainable food going 'global', thereby putting the spotlight on the relationships between non-contiguous places. In short, producer choices for markets in one location were argued to be causing ecological degradation in another distant one (Swanson, 2015).[3]

Against these background debates, in this Chapter, I first set out what the issue of sustainability as a food governance problem 'could be' as discussed in the scientific literature on fish farming aquaculture. I then proceed to examine in each sector and territory the institutionalisation of sustainability as a food governance problem. Applying my framework, in each case I will show how actor political struggles over nature–society interdependencies combined with struggles over the frontier politics of interdependence and territory, regulation and knowledge have determined the choices resulting in particular 'settlements' of sustainability interdependence.

Sustainability as a food governance problem

The aquaculture literature addresses various issues associated with sustainability as a food governance problem from different angles and from within different disciplines, especially economics, geography and feed nutritional science. I have grouped scientific debates under three headings: (i) a literature on the marketing of farmed fish products addressing their biophysical attributes; (ii) a literature on 'green consumerism'; and (iii) a literature on product quality and the substitution of FM and FO in fish feed diets.

Initial discussions around sustainability marketing strategies for farmed fish began at the end of the 1990s and, as argued by Young *et al.* (1999), at a point when different farmed fish markets were reaching their maturity and farmed fish product prices were beginning to fall. In response, these scholars made a call for new knowledge on 'consumers' tastes and attitudes towards aquaculture products' (Young *et al.*, 1999: 7). A first strand of this research analysed the marketing of farmed fish products with specific attention being paid to biophysical issues. Addressing what researchers defined as the 'environmental attributes' of farmed products, this work focused on the local market relationship of farmed fish products to their wild caught counterparts. This was an important relationship to grasp, they argued, not least because of increasing consumer 'connotations of seafood as a healthy alternative protein source to red meat' (Young *et al.*, 1999: 10).

Accordingly, the effective marketing of farmed products could be achieved through the stressing of their 'naturalness' and 'similarity to wild captured stocks' (Young *et al.*, 1999: 10). This was not to suggest that farmed fish could necessarily substitute for wild fish. For example, in their literature review on the interactions of fisheries and aquaculture at a global scale, Natale *et al.* (2013) found little evidence of market integration: in fact, producers were creating new markets for their farmed products and commanding different prices. This was also confirmed in research on local markets. For example, analysis of Spanish seabream seafood markets has shown that wild and farmed seabream are complementary goods (Fernández-Polanco and Luna, 2010).

As a result, Fernández-Polanco and Luna argued that producers could 'take advantage of the market appraisal of their wild equivalents' in marketing strategies (2010: 44). Work to render the product as natural as possible, for example through the selling of the fish whole or gutted, they argued, could be to producers' marketing advantage. In addition, producers could promote a consistency of supply which farmed products could offer over their wild caught counterparts. Work on naturalism was important, they argued, because 'technology ... [was] a potential source of risk for concerned consumers' (Fernández-Polanco and Luna, 2012: 23). In this vein, they also posited that in terms of consumer perceptions, traditional food stores were more likely to stock top-quality products marketed as being 'close to nature', whilst supermarkets were more likely to stock a standard product (Fernández-Polanco and Luna, 2012: 26), thus encouraging multiple contracts between producers and retailers.

Running parallel to this research on environmental attributes in terms of the 'naturalness' of products, a second strand of research has focused specifically on the theme of 'green consumerism' in product differentiation (Young *et al.*, 1999). As noted by Young *et al.*, at the end of the 1990s not many fish farming companies had engaged in product differentiation or sustainability standardisation practices (1999). They argued that this could be an important marketing strategy to be pursued.

Drawing on lessons learned from wild seafood markets also turning towards green consumerism, it was argued that work to give market value to the positive environmental attributes of farmed fish from this perspective would require communicating to consumers more complex information concerning farming processes. This was because green consumerism required going beyond the product to considerations of how it was grown, since 'perceptions about seafood harvesting methods are as important as those about the product itself' (Fernández-Polanco and Luna, 2012: 23). This could be facilitated by product eco-labelling (Young *et al.*, 1999) which would act as a 'symbol for exposing differences between production practices according to their sustainability' (Whitmarsh and Wattage, 2006: 117). Additional research conducted on seafood had shown that environmental labels could be more useful than mere claims of product quality in commanding premium prices (Jaffry *et al.*, 2004).

Following these kinds of arguments, research examined different consumer communities' 'willingness to pay' for labels for farmed fish and concluded that consumers were indeed willing to do so (see e.g. Olesen *et al.*, 2010 on Norwegian salmon; Whitmarsh and Wattage, 2006 on Scottish salmon). Consequently, in these cases at least, green consumerism facilitated through labelling potentially produced

> a financial benefit, in the form of higher prices, from product differentiation based on environmental attributes.
>
> (Whitmarsh and Wattage, 2006: 117)

In these articles focusing on how to benefit in the market place through product differentiation and eco-labelling, communicating positive environmental impacts to consumers, the discursive and knowledge processes involved in distinguishing a 'sustainably farmed fish' from a 'farmed fish' was not the main focus. This issue has, however, been extensively treated within human geography by Mansfield (2003). In the case of US farmed catfish, she revealed the political economic conflicts which arose when actors sought to distinguish certain farmed fish from others on the basis of their biophysical and farming processes. She demonstrated how making distinctions by applying these criteria not only created political competition amongst fish farmers and fishers within the domestic US seafood market, but also mobilised fish farmers to 'make a protectionist argument' seeking to limit domestic market access for 'foreign' farmed fish products (Mansfield, 2003: 332–333).

Engaging in 'green consumerism' in this case not only resulted in public and private actors distinguishing products on the basis of their biophysical process, but also on the basis of their food production geographical space. Consequently, for Mansfield, the issue associated with labelling farmed fish as sustainability food ultimately became a geographical one (2003: 340), in which sustainability-related arguments were used as a weapon. This was also shown to be the case in political economy research into European debates on the sustainability of farmed fish and also over (imported non-EU) catfish (pangasius), whereby product differentiation of farmed fish in the name of sustainability was raised as a potential EU trade issue. Indeed at times this has involved 'claims and counter claims' about the sustainability of products in a 'battlefield of knowledge' over farming practices (Bush and Duijf, 2011: 185, 187).[4]

Third, coming at the question of farmed fish as a sustainability food governance problem from quite a different angle, a body of work has grown up around product quality and the substitution of FM and FO in fish feeds (Natale *et al.*, 2013). Ever since the BSE ('mad cow disease') and pig dioxin food safety crises in the 1980s (in the UK and Belgium), feeds have been identified as being intimately linked to social constructions of food product quality (Carter, 2012, 2015a). Yet whereas fish feeds had already been associated with product quality defined in terms of safety, the article by Naylor *et al.* (2000) (discussed above) strongly associated feeds with product quality defined in terms of sustainability.[5] This raised two potential challenges. First, to lower the levels of FM and FO in feeds through substituting them with other ingredients; and second, to improve the management of feed fisheries (Carter, 2015a). For, as argued later in a follow-up article by Naylor *et al.*, only through the finding of alternatives will 'a consensus … emerge that aquaculture is aiding the ocean and not depleting it'. Consequently, 'effective regulation of global forage fish supplies … is critical' (Naylor *et al.*, 2009: 15109).

In response, a further literature has grown up around substitution, documenting trends in the continued use of fish meal (FM) and fish oil (FO). For example, Tacon and Metian found a downward trend in the use of FM and FO (2009) and Natale *et al.* (2013) confirmed this downward trend in 2013. Additionally, scholars have debated the calculations behind the Fish In, Fish Out (FIFO) equation initially politicised by Tacon and Metian (2008). Tacon and Metian had calculated the amount of FM and FO needed to produce a kilo of farmed salmon and concluded this gave a ratio of 5:1. Yet the calculations behind their equation have been hotly disputed (Byelashov and Griffin, 2014; Jackson, 2007; Kaushik and Troell, 2010). These technical discussions notwithstanding, the ratio has remained a powerful symbol used both by those attacking aquaculture and those seeking to defend its adaptation to feed sourcing issues (Byelashov and Griffin, 2014).

Second, an extant literature and research programme on fish nutrition has consolidated to address the issue of FM and FO substitution. This research has attested that substitution is not a straightforward technical issue since marine resources are vital components of fish diets. Not only does FM contain essential proteins for fish growth, but FO contains long chain fatty acids (Omega 3 and 6) whose trace elements provide an essential nutritional value of farmed fish for human mental health (Crawford 2010).

Many research articles have been published on these very questions. For example, Drakeford and Pascoe (2008) have worked on vegetable substitutes for marine ingredients; Tusche *et al.* (2011) have shown the benefits of organic potato protein for trout; Bendiksen *et al.* (2011) have argued that it is easier to replace FM than FO in salmon diets; and Gillund and Myhr (2010) have carried out research on deliberative assessments of consumers' values related to the replacement of FM and FO. These and other works have revealed the tensions between sustainability of product quality through substitution and other attributes of product quality such as fish and human health. This tension has crystallised in the EU over one potential substitute, namely, Processed Animal Proteins (PAPs) (Carter, 2015b; Naylor *et al.*, 2009).

PAPs are rendered products and are sourced from abattoir or catering waste of pigs and poultry which has been certified as 'fit for human consumption'.

Fish nutritionists have argued that PAPs can be very effective substitutes as they contain important amino acids and are nutritionally valuable, especially for fish (Bureau, 2006). It is argued not only that PAPs can be used to replace fish proteins in diets, but also that they are more effective sustainable protein replacements than vegetable ones since they are easily converted by fish, have nutritional value and are recycled products.

However, the use of PAPs in feeds is controversial. For many years their use was banned in EU fish farming production under EU food safety regulation.[6] This changed in July 2012 when they were re-authorised by the EU Standing Committee on the Food Chain and Animal Health (Carter, 2015b). However, this decision was met by a hostile public reaction in some countries. A central concern was that feeding PAPs to fish was against nature: 'You have never seen a fish eat a pig' (José Bové, cited by Mauriac 2013). Ultimately, therefore, debates over feed substitution with PAPs echo earlier discussions around the 'naturalness' of farmed fish products in marketing strategies. They thus confirm the centrality of 'the theme of original nature vs. the copy' (Mansfield, 2003: 338) in constructions of sustainability as a food governance problem for farmed fish.

In summary, there are a number of issues which 'could be' associated with sustainability as a food governance problem for farmed fish. These include issues of biophysical equivalences between wild and farmed products; green consumerism, product differentiation and eco-labelling; and product quality and feeds. In the rest of this chapter I examine how actors involved in the markets of Scottish salmon, Aquitaine trout and Greek seabass and seabream have actually perceived this issue and worked institutionally on this particular question.

Scottish salmon and sustainability as a food governance problem

Overview of the changing sustainability narratives

There are two central elements in the story about sustainability as a food governance problem for Scottish salmon. First, the governing of Scottish salmon as food has increasingly been brought about through private self-regulation in a logic of 'freedom to compete' (Smith, 2016). Second, the contents of this form of governance have been altered. A key change has been from conceptualising sustainability in terms of local farm impacts to its conceptualisation in terms of the interdependent ecosystem impacts of non-contiguous places.

Initial constructions of Scottish salmon as a food product were coherent with a social meaning of sustainability which depoliticised ecology–economy interdependencies (see Chapters 4 and 5). In the beginning, Scottish salmon was primarily sold as a commodity product onto a generic UK market rivalling Norwegian salmon, whereby Scottish companies held multiple contracts with supermarkets competing on quality and price.[7] The only product differentiation was for salmon sold on the French market under the product quality label 'Label Rouge' (LR). This choice of label, however, was not linked to sustainability issues:

> [We] saw it as a niche market and good for business ... good for image too and getting great prices.
>
> (Interview: Scottish salmon producer, 2010)[8]

Otherwise in the early years of production and emerging markets, the environmental attributes of farmed salmon were not politicised in marketing strategies. Rather, because farmed salmon products could contribute to seafood markets by compensating for a decline in wild capture fisheries, salmon could be considered as 'naturally' sustainable.

In the mid-2000s, however, different types of political work undertaken by different groups of collective private actors 'outside' the sustainability institution had the effect of disrupting this social construction of farmed salmon food products as being naturally sustainable. In particular, the issue of salmon feeds was raised as critical in EA constructions of sustainability. Up until now, fish feeds had been associated primarily with animal nutrition, not with sustainability:

> We were doing the one product range ... life was very easy then, people weren't bothered about where the FM and FO came from, sustainability [of the feeds] wasn't talked about ... all people were really concerned about was the price of feed and the kind of performance they could get from it.
>
> (Interview: Scottish feed company, 2010)

This was to change, however. First, in 2004, US and Canadian scientists published an article in *Science* claiming that Scottish salmon products contained high levels of potential cancer-causing contaminants which put consumers' health at risk if they ate more than 'one-half meal of salmon per month' (Hites *et al.*, 2004: 228). The researchers claimed that the levels of these contaminants in the Scottish farmed salmon were much higher than in the wild salmon and further that they came about due to the fish feeds. This was because the FM and FO used in Scottish salmon diets were, they claimed, frequently sourced from NE Atlantic pelagic feed fisheries, themselves heavy in PCBs (polychlorinated biphenyls – organic chlorine compounds).

In response to this article, a number of different UK and EU actors swiftly joined together in a damage limitation exercise led by the SG, the salmon producers' organisation, the European Commission and the UK Food Standard's Agency. Their collective response sought to depoliticise the issue: statements were made that Scottish farmed salmon had not been compared to Scottish wild salmon, but to Pacific wild salmon which have a different diet; that the levels of PCBs in the salmon were below the safety thresholds set by the European Commission and the World Health Organisation; and that eating a fish-based diet was still healthier than eating a red meat-based one.[9]

This was followed later on in 2004 by the publication of a joint e-NGO report politicising the methods used by Scottish feed companies and producers to judge the sustainability of the feed fisheries from which they sourced their marine ingredients (Huntington, 2004). These, the report argued, were missing wider ecosystem impacts that should otherwise be captured (Huntington 2004). The presentation of arguments to place the feed companies' FM and FO sourcing policies in the public domain continued into 2005 with a campaign by Greenpeace to mediatise UK supermarkets' seafood sourcing policies. Ranking UK supermarkets' policies against ecosystem approaches, Greenpeace 'named and shamed' retailers, claiming that many of their policies were unsustainable (Greenpeace, 2005).

The institutional response to these accusations against Scottish salmon as a safe and sustainable food product has been political work renewing partnerships between supermarkets, producers and feed companies[10] as well as the creation of new alliances with e-NGOs, to set standards for differentiated products:

> The recent strategy was driven out of the PCB dioxin scare in 2004 and none of the UK retailers could say 'we are ok' in that 'we have a differentiated offer'. So that is what drove the responsibly sourced offer.
>
> (Interview: UK supermarket, 2012)

Through their voluntary self-regulation and new standard setting, these private actors have collectively brought about a re-institutionalisation of sustainability as a food governance problem for salmon. Business to business commercial relationships between producers and supermarkets would no longer be governed by the mantra: 'here is the contract, there is the price, this is the quality we want' (Interview, Salmon producers' association, 2012). On the contrary, instead of multiple contracts between producers and supermarkets selling a commodity product, exclusive contracts have been drawn up between supermarkets, producers and feed companies in which differentiated products must meet sustainability standards as agreed by all parties to the contract:

> It is about 'we're going to invest in the farming *method*, the standards that are applied to it, the message, the provenance – everything about it'. It's all about more than just the contractual need for the fish to be on the market.
>
> (Interview: Salmon producers' association, 2012)

The social meanings of sustainability governing these contracts were no longer ones which solely depoliticised ecology–economy interdependencies. Rather, actor understandings of sustainability shifted towards recognition of the inter-relations between socio-ecosystems in non-contiguous places. Private standards governing product specification now not only expected the substitution of FM and FO, but also that fish feed ingredients (whether marine ingredients or vegetable ingredients) be sustainably sourced. These new collective understandings thus institutionalised anthropocentric eco-system approaches by problematising ecology–society interdependencies within private self-regulation. Yet, as we will see, the individualised character of the business to business contracts which would now structure the market place would result in these interdependencies being institutionalised within a 'logic of freedom to compete' (Smith, 2016).[11]

Unpacking interdependencies

Shifts in the social meanings of the sustainability of environmental attributes of products and their shadow ecologies were brought about through institutional and political work whereby the politics of nature–society interdependencies have been embroiled in a frontier politics of interdependence and territory, regulation and knowledge.

Territory: Unlike over fish farm/environment interactions and access to fish farm sites, Scottish devolution's territorial resources have not been mobilised by public actors to govern the marketing of food products or indeed the regulation of fish feed contents. On the one hand, the contents of feeds are an exclusive EU competence under EU food safety policy and Scottish actors have not sought to politicise the frontiers of EU jurisdiction on this matter.[12] On the other hand, the marketing of food products in general has been 'left to the market' by Scottish public actors.[13]

> Scottish Ministerial working groups are having very little direct impact on the feed industry. We are impacted by EU rules and haven't seen any dramatic change in its interpretation when applied locally.
> (Interview: Scottish feed company, 2012)

> The Scottish Government don't see it as their issue.
> (Interview: Scottish feed company, 2012)

However, the absence of a Scottish public policy on this issue does not denote an absence of political work on the frontiers of territory. Collective private actors have engaged strongly over territorial politics defining farmed fish products as sustainable food. This political work has been carried out on a number of complementary fronts.

First, within Scotland, private companies have worked collectively through their producers' organisation to associate salmon farmed in Scotland with Scottish geographical territory. This led in 2004 to the obtaining of an EU Geographical Product Identification (GPI) for 'Scottish' salmon. Since then, and following the publication of the Hites *et al.* (2004) article, Scottish actors have engaged in extensive voluntary private self-regulation (see below). In this manner, they have defined the 'sustainability' of 'Scottish' salmon as a food product.

Although there are different degrees of emphasis depending on the private product specification in question (see below), in general, sustainability in Scottish salmon as a food product has been defined both in terms of salmon farm/environment interactions as well as concerning the substitution of FM and FO and the sourcing practice of fish feed ingredients (i.e. both local and distant ecosystem impacts of salmon farming). This has been accompanied by different strategies of institutional work at the frontiers of territorial action. First, even if the Scottish working group on image and marketing[14] has not brought feeds directly within its remit, it has developed strategies for the local territorial insertion of Scottish salmon as a 'healthy' food product for Scottish consumers. This has included, for example, working with Scottish schools to encourage children to eat salmon:

> The salmon industry produced a package which taught school pupils about ... salmon production and asked them to design a meal involving salmon ... asked them to market the meal and then eventually the meal made its way onto the school dinner menus in the Highlands.
>
> (Interview: Scottish public actor, 2012)

Second, Scottish private actors have engaged in UK and EU-wide discussions seeking to bring the contents of UK supermarket policies and EU public rules on food safety in line with their own definitions.[15] One such issue has been over PAPs. According to Scottish actors, PAPs constitute sustainable substitutes for marine ingredients not only because they reduce pressure on wild fisheries and are nutritionally useful for fish, but also because their use can valorise animal by-products which would otherwise be wasted and thrown away:[16]

> Nutritionally ... fantastic ... very high in protein, very rich in amino acid which isn't available in other raw materials.... It is a raw material which is a by-product from another industry ... you are not getting the full value from them when they are going into landfill or energy production ...
>
> (Interview: Scottish feed company, 2012)

Scottish producers and feed companies have both lobbied UK supermarkets and engaged in EU-wide arenas to bring both private and public rules on PAPs in line with their own interests. Yet, whereas thus far they have been unsuccessful in changing UK supermarket policies on this particular issue (see below), feed companies have successfully engaged via their EU-wide federation (FEFAC) to reverse EU rules prohibiting the use of PAPs in fish feeds (which happened in July 2012) (Carter, 2015a; 2015b).[17]

Otherwise, a central element of Scottish private regulation on salmon as a sustainable product has been its focus on the provenance of raw materials used in feeds, whether FM and FO or other ingredients, including vegetable sources:

> I guess the most important step is traceability – you have to know what you are actually buying and where it has originated from.
>
> (Interview: Scottish feed company, 2012)

Third, collective private Scottish actors have also worked institutionally on the global scale to influence the contents of sustainability being set in global standards. In this work, some elements of the Scottish definition of sustainability within an ecosystem approach have come into conflict with the negotiated global definition. This is because, at the end of the 2000s, WWF launched multi-stakeholder Aquaculture Dialogues to set global sustainability ecosystem standards for aquaculture products leading to an Aquaculture Stewardship Certification (ASC).[18] One of these dialogues was on salmon, led by a global partnership of leading salmon farming companies (such as Marine Harvest) and e-NGOs, in which Scottish WWF played a critical role.[19]

Whereas initially Scottish producers were not involved in these global discussions, they became increasingly engaged. As argued by Havice and Ilas (2015), the contents of definitions and rules set for the ASC were the result of power struggles and negotiation between different groups of private actors representing economic, ecological and territorial interests, some of which ultimately won out over others.[20] In the case of the salmon dialogue, Scottish salmon producers considered the contents of the standards to be very much influenced by others' definitions of sustainability, especially North American and Norwegian constructions:

> But I also noticed at the time that it was very much America's; it was very focused on non-European aquaculture.
>
> (Interview: Salmon producers' representative, 2012)

This had important implications, for example, for the institutionalisation of global social meanings of the ecological impacts of freshwater farming on biodiversity. According to the standard, advertising its salmon as ASC accredited (and hence sustainable) would commit a business to cease farming juvenile salmon in open water. Scottish producers have therefore found themselves confronted with a global standard which is in contradiction to the Scottish one already codified:

> The main one is that we produce smolts – juveniles – in fresh water systems. And we have evolved a [Scottish] standard of regulation for that, which in our view is entirely sustainable, it is a very, very high standard. Because the Norwegians have never really had to do that and technology

has moved on and recirculation hatchery technology means you can actually invest large amounts of money in putting things on to tank-based systems in-land, the view is that creates barriers and therefore there is less risk with that stage of the process of production. So the [ASC] standard has been set on everything being based on land-based facilities and nothing in fresh water. So effectively the very first stage of our entire production process excludes us from the [ASC] standard.

(Interview: Salmon producers' representative, 2012)

Just thinking about the new Aquaculture Stewardship Council certificate. One of the things in there is if you have freshwater loch sites, that you will not have them in five years' time. But who says they're wrong? We have a number of freshwater loch sites, … well why is that wrong? Because we have always found that fish that come from a loch site are better at sea. They perform better. Ok you don't have the containment the same as you have in a tank, but you contain them at sea and you just need to transpose that into freshwater and that's how we do it.

(Interview: Scottish farm site manager, 2012)

Another issue was over the lethal control of predators such as seals:

The standard does not allow for the lethal control of predators such as seals. We in Scotland have a large population of protected seal species that exist within the vicinity of our farms. I am not saying that is unique, because in Canada they have seals, less so in Norway … but we have 80 per cent of the entire world population of the common seal in Scotland in the same place that we farm our fish; and that is a protected species – not protected, but we are constrained in what we can do with that. But if we have a policy that states 'we are not going to give you accreditation unless you can absolutely guarantee that you will not dispose of a seal that has got into your cages', I don't think we can do that either. So unless we find a means of exclusion that actually works and we can guarantee that a seal cannot get to our stocks, then we are not going to be able to sign up to that standard.

(Interview: Salmon producers' representative, 2012)

Thus far, the contents of these ASC rules have been constructed by some Scottish actors as being more connected to the building of other territories' definitions of sustainable food than their own in the name of a global standard. However, this is not a straightforward situation. This is because first, not all salmon companies farm their juveniles in open water (and hence could meet that part of the standard) and second, because although it is possible to identify some common elements of Scottish political work on sustainability as a food governance problem, nonetheless private self-regulation has been conducted within a logic of competition. This means that in the future companies in Scotland might start to compete with one another on this very issue (e.g. through applying for ASC accreditation).

Regulation: As can already be surmised, political work on the frontiers of territory has been inextricably interwoven with political work on the interface of public and private regulation. Up until the mid-2000s, even if individual companies adopted in-house policies for managerial standard setting, there was very little 'external recognition impact' of salmon products (Young *et al.*, 1999). Indeed, the environmental attributes of products were not explicitly communicated in marketing strategies.

The shift in Scotland towards 'green consumerism' in the mid-2000s was accompanied by extensive voluntary self-regulation and, in the area of fish feeds, policy capture by economic interests. This came about through the opening up of discussions between supermarkets, producers, feed companies and e-NGOs on product specification. Of course, in the explicit selling of farmed fish within a narrative of sustainability, criteria need to be set to specify what a 'sustainable' farmed fish means from this perspective and how this will be judged. Criteria are required irrespective of whether 'sustainability' is being expressed through a label such as 'organic', through a private brand or through a general supermarket message to consumers.[21]

On the one hand, one set of criteria for judging sustainable salmon has been over farming methods and in particular over local fish farm/environment interactions. In some cases, supermarket sustainable sourcing policies have been based upon an understanding that Scottish public environmental regulation is sufficient to meet their objectives concerning local ecosystem impacts:

> Clearly there is a governance regime in Scotland covering the salmon farming industry. And from our perspective we are not asking our supply base to do anything significantly different in terms of the basic regulatory regime ... so things like discharge consents, that is the framework for operation in Scotland and I think it is a very robust framework from that point of view. So we wouldn't dictate things like the number of cages on a site. That is not our expertise. That is planning authorities and SEPA [Scottish Environmental Protection Agency]. They are the custodians of that.
>
> (Interview: UK supermarket, 2012)

In other cases, additional local farm/environment conditions have been specified over and above those set in public regulation. These have been contained either within the rules of the label[22] or in the contracts negotiated between producers and supermarkets.

Consequently, there are different relationships in evidence between public and private regulation governing local ecosystem impacts. Some private self-regulation sets additional conditions which supplements public regulation, whereas others do not. By contrast, on the issue of environmental impacts on distant socio-ecosystems, private self-regulation clearly substitutes for public regulation. Regulating this issue has brought collective private actors into political discussions both over the replacement of FM and FO with other ingredients, as well as over the demonstration of sustainability in the sourcing of feed ingredients.[23] Concerning the replacement of FM and FO, this problem has been worked upon to the point that it now commands widespread acceptance amongst actors across the industry. Indeed, many actors have been very critical of Label Rouge in this regard, constructing it as unsustainable because it demands high levels of FM and FO for fish feed diets:

> Label Rouge is pulling in the opposite direction on sustainability, if you want to have less impact on FM and FO.
>
> (Interview: Scottish feed company, 2010)

> For Label Rouge – this is the big problem – Label Rouge is not about sustainability. Label Rouge is purely about the eating experience ... So for Scottish Label Rouge, it is 70 per cent marine contents, 45 per cent FM and the rest FO which is ludicrous ... it makes my blood boil.
>
> (Interview: Scottish feed company, 2012)

Actors have therefore worked to reduce FIFO (Fish In, Fish Out) ratios in feed recipes and replace these ingredients with other ingredients. A central argument has been that any available FO on the global market should be used as far as possible for fish feeds, as distinct from other uses:

> Back in the 50s and 60s digestive biscuits were full of fish oils, which is a very poor use of a valuable resource. But now we like to make sure that we use every drop of fish oil and that we use it in fish.
>
> (Interview: Salmon producers' representative, 2012)

However, having agreed that FM and FO replacement was necessary, and having reduced FIFO ratios in salmon feed diets, the critical issue which has emerged has been over designing fish feed recipes to ensure that a salmon product meets several market requirements simultaneously, i.e. in terms of human health, sustainability, and fish health.[24]

In these debates on product quality in line with different representations of that quality, there have been conflicts. For example, although for the most part, marine ingredients have been replaced with vegetable ingredients, there are limits to these replacements. Indeed, other ingredients potentially offer greater potential, both on grounds of fish performance as well as on grounds of sustainability.

One such ingredient is PAPs, yet producers, feed companies and supermarkets have disagreed over the use of PAPs in diets. UK supermarkets will not buy farmed fish fed on PAPs which they do not consider to be in line with consumer demand. On the other hand, producers and feed companies

and some e-NGOs consider PAPs to be nutritionally valuable replacements which satisfy the criterion of sustainability in the market place.
There have also been disagreements over the use of GM (genetically
modified) oils and cereals as replacements. This has put Scottish salmon
products in a competitive situation compared with other salmon products
(e.g. from Chile), where the fish have been fed diets using a wider range
of ingredients:

> The list of options becomes smaller as you move into the UK.
>
> (Interview: Scottish feed company, 2010)

Therefore for these reasons as well, alongside responding to food safety[25] and
sustainability concerns, producers, feed companies and retailers have worked
to add value to products through working on the provenance of the raw
materials used and in some cases making bespoke diets for consumers.

Indeed, a critical feature of Scottish private self-regulation and standardisation has been its setting of in-house rating systems of (far-away) socio-
ecosystem impacts following the harvesting of raw materials. Frequently,
these have been set in addition to global standards and EU public rules on
traceability, thus creating another set of private institutions governing sourcing practices.[26] This has happened when global standards have been supplemented through the development of sustainability sourcing policies within
which private standards and rating systems have been set in partnership
between supermarkets, feed companies, e-NGOs and producers. This has
been facilitated by the creation of new posts in UK supermarkets for sustainable sourcing policy officers and/or new posts created within e-NGOs to
generate expertise on these specific questions.

Finally, it has not just been the provenance of marine ingredients which
has been at issue, but also the provenance of vegetable ingredients. For
example, feed companies have policies only to source soya from Brazil (on
the grounds that it is GM-free) and from companies which are members of
the Roundtable for Responsible Soya Production. This has resulted in a
detailed work of refinement over feed ingredient criteria within competitive
market strategies:

> [Now we] have gone more to provenance as opposed to just the com-
> position sustainability. Customers don't just want FO, but anchovy oil or
> blue whiting oil. For example, palm oil is excluded because it is not con-
> sidered sustainable and customers won't buy fish fed on palm oil.
>
> (Interview: Scottish feed company, 2012)

In the case of both FM/FO substitution and the provenance of fish feed
ingredients, therefore, voluntary private self-regulation has substituted for
public rules, simultaneously regulating new markets following a logic of competitiveness (Borraz, 2007).

<u>Knowledge</u>: Finally, interwoven with these politics of interdependence of territory and regulation has been an important politics of different knowledge types supporting sustainability marketing choices. On the one hand, feed nutritional knowledge is required to support choices over appropriate ingredients and recipes. To this end, feed companies have generated extensive nutritional knowledge through their R&D departments. Additionally, working collectively through SARF (the Scottish Aquaculture Research Forum), public and private actors have funded projects looking for alternate replacements for FM and FO. On the other hand, making choices over feed ingredients has resulted in both the mobilisation of existing information and the generation of a whole new kind of knowledge, ranking feed fisheries on grounds of sustainability supporting private regulation:

> We have a means of measuring them or at least expressing them.
> (Interview: Scottish feed company, 2010)

> So we have a sustainability rating system for all the farmed fish … that we sell.
> (Interview: UK supermarket, 2012)

> We spend a lot of time talking to the feed companies about the ingredients that go into diets. With one of our major customers, we sit down with them on a regular basis and review the species of fish that go into our diets. We have the red, amber, green system. Looking for green, will take amber if looking positive, won't take red. And that is done on a regular basis.
> (Interview: Scottish salmon producer, 2010)

This has resulted in disagreements between e-NGOs and private economic actors over the way in which the distant environmental impacts of raw materials are being measured and whether these are being measured within ecosystem approaches. Environmental NGOs have positioned themselves institutionally on this question, as holders of alternative ecosystem knowledge on feed fisheries. This has given them access to meetings with producers, feed companies and supermarkets setting standards. In some cases, e-NGOs have thus been heavily involved in grading fisheries, contributing to rating systems. However, having access has not always translated into influence over ultimate choices made:

> We are always aware of what environmental NGOs are saying. But we have got our policies and we don't always agree.… They have got their rating system; we have got ours. Sometimes they are different.
> (Interview: UK supermarket, 2012)

In this political work at the frontier of interdependence and knowledge, tensions within the interdependence have therefore both been narrowed down

Table 6.1 Summary of political work over sustainability as a food governance problem for Scottish salmon

Scottish salmon			
Social constructions of sustainability	Attacks on sustainability as de-politicised local ecology–economy interdependencies lead to a re-institutionalisation implementing ecosystem approaches which recognise inter-relations between socio-ecosystems in non-contiguous places within a logic of freedom to compete		
Frontier politics of polity interdependencies	Territory	Regulation	Knowledge
	No separation of Scottish public approach; EU public rules on food safety dominate; Private rules define social meanings of 'Scottish' sustainable salmon as food and private actors engage in multi-positioning strategies	Co-regulation governing sustainability marketing criteria on local environmental impacts; Policy capture by economic interests on sustainability marketing criteria on fish feeds; Conflicts over PAPs as replacements	In-house company knowledge dominates regulatory choices on product specification; Political usage of knowledge by e-NGOs to institutionally position themselves as experts and gain access to discussions over product specification

to controversies over food safety and opened up to debates defining sustainability in a logic of freedom to compete (especially in the UK market).

Aquitaine trout and sustainability as a food governance problem

Overview of the changing sustainability narratives

Debates on sustainability as a food governance problem for Aquitaine trout reveal two main developments which have dominated its government. First, in a situation of industrial decline, sustainability has increasingly been couched in terms of equality through private self-regulation. Second, as was the case for Scottish salmon, understandings of ecosystem impacts have not been limited to those 'at home' but have been extended to include those in 'far flung' fish feed production territories.

Since its early establishment in the 1980s, Aquitaine trout has experienced a variety of routes to market. Larger companies, such as Aqualande or Viviers de France, have changed their marketing strategies quite radically over the years through product diversification and differentiation,[27] selling mainly to supermarkets. For smaller enterprises, a variety of practices can be observed. For example, in the Pyrenees and in the Basque country, as well as selling processed products to supermarkets, companies sell fresh whole fish to local fishmongers or restaurants, whilst others still have engaged in product diversification selling processed niche products such as. trout rillettes in local markets and trade fairs. These differences notwithstanding, it is still possible to identify common institutionalisations of sustainability governing the sale of Aquitaine trout, not least because producers in all locations engage in common collective associations[28] which have worked on its very meaning in alliance with supermarkets, processors and WWF at a French-wide scale.[29]

What we can observe in the case of Aquitaine trout is an institutionalisation of sustainability of trout as a food product problematising ecology–society interdependencies and which is increasingly linked to the sustainability of 'the homeland'. This can be understood especially in the context of the decline in the growth of this industry and perceived threats from ecocentric applications of ecosystem approaches governing both farm/environment interactions and access to farm sites (see Chapters 4 and 5).[30]

This problematisation of trout as a food product did not happen all at once, however. Initially, Aquitaine trout was sold as an unprocessed commodity product in the French market place.[31] Very quickly, large companies realised that they needed to compete with Norwegian salmon to grow their business. This necessitated growing larger fish in order to diversify the product to sell trout steaks and smoked trout.[32] For example, in the cooperative Aqualande this initially led to product branding which imitated Norwegian products. The company created a brand name 'Landvika' being 'Land' for 'les Landes', and 'vika' to make the product sound Nordic (interviews).

However, as trout production in France began to decline overall, marketing strategies changed, increasingly becoming coherent with social meanings of sustainability as territorially embedded anthropocentric ecosystem approaches:

> We separated from the Nordic universe and are now in the French universe.
>
> (Interview: French trout producer, 2012)[33]

This resulted in the creation of new and replacement brand names; again, for example, in the case of Aqualande, the brand name OVIVE was created (i.e. 'eau-vive', or 'living water' in English):

> It is a brand where we really apply sustainable development, the original territorial aspects of French trout, but also Pyrenean trout, Basque trout, Aquitaine trout through specifying its health aspects, Omega 3, because trout is less fatty than salmon and naturally rich in Omega 3. And we make a final important point, we accentuate this point, that in buying the brand OVIVE you are preserving local employment.
>
> (Interview: French trout producer, 2012)[34]

Indeed, partly in response to the decline in the growth of industry, a distinguishing feature of French trout has been extensive collective private self-regulation led by the French-wide inter-professional association CIPA (Comité Interprofessionnel des Produits de l'Aquaculture) and focusing on common sustainability marketing strategies. Rather than seeking to create competition amongst producers (as was the case for Scottish salmon), one aim has been to establish a collective private brand called 'Charte qualité et aquaculture de nos régions', available to any producer or processor signing up to the collectively negotiated quality standard condensed by the brand. Behind these marketing initiatives has been a concern to work to protect sustainable French trout products in competition with other products – e.g. not only Norwegian salmon, but also imported non-EU farmed fish products such as pangasius.

Whereas initially the environmental attributes of product quality vocalised in marketing strategies were linked to water quality and sustainable farming methods, increasingly actors began to pay attention to fish feeds and the use of marine ingredients (FM and FO).[35] Indeed, already in the product specifications developed in the mid-1990s for the supermarket Carrefour's own quality brand for trout ('filière qualité'), feed ingredients' sustainability was a key factor in defining product quality.

In 2011, these questions were tackled head on when CIPA, working in alliance with WWF France, took the initiative to negotiate a single product specification for contracts between French trout farmers and supermarkets. Central to definitions of sustainability behind the product specification were discussions over feeds. Not only was it agreed that product specification should include rules on percentages of FM and FO in feed diets, but also that marine ingredients had to be sustainably sourced. In this manner, continued institutional work on marketing has updated social meanings of the sustainability of trout as a food product. As a result, social constructions of sustainability as homeland have been complemented by new social representations of wider socio-ecosystem effects between Aquitaine and other distant regions.

Unpacking interdependencies

Whereas in Scotland, I identified anthropocentric ecosystem approaches within a 'logic of freedom to compete', in Aquitaine trout by contrast I have identified anthropocentric ecosystem approaches within a 'logic of equality' (Smith, 2016). These have been stabilised through political work over nature–society interdependencies in connection with political work on the politics of interdependence of territory, regulation and knowledge.

Territory: As we saw in the case of Scottish salmon, Aquitaine trout as a food product is governed through EU public rules and standards on food safety. For the most part, French public actors have not sought to challenge EU political authority on this issue.[36] Instead, political work territorialising farmed trout as a food product has advanced on two other fronts.

First, collective private actors have sought to institutionalise strong local identifications of trout farmed within Aquitaine through making connections between product quality and local territorial identities and geographical origins, e.g. les Landes, the Pyrenees and the Basque country. This has been codified when producers have developed collective private brands capable of being translated territorially,[37] or when actors have made preparations for a local protected geographical indicator PGI, as has been the case for producers in the Basque country. In these initiatives, producers have constructed the quality of the produce in line with where it has been produced:

> We will justify our product's quality through demonstrating that we are integrated into the milieu and that our activity is not delocalisable.
>
> (Interview: Trout producers' representative, 2012)[38]

In their institutional work on markets, producers have linked the sustainability value of the product with the sustainability value of the territory and made distinctions between what they have defined as a difference between an organic and a sustainable product:

> Organic and sustainable, these are two different concepts for me. Organic is a way of producing which of course respects the environment. Sustainability is a wide concept, because sustainability is not just about the environment.
>
> (Interview: French feed company, 2012)

Marketing strategies premised upon these kinds of understandings have been accompanied by an important investment in local markets. This investment has been congruent with general shifts in French society responding to global change and the setting of regional public policies to encourage the establishment of local food supply chains (Naves, 2016):

> I think that there has been recent change in favour of local consumption. Local heritages are coming back.
>
> (Interview: French trout producer, 2012)[39]

On the one hand, trout products have been expressly marketed as 'local' products in the supermarkets. On the other hand, producers have developed marketing strategies of creating local markets and selling fish to local shops, restaurants, school canteen platforms and other outlets. In this way they are sustaining the local economy (for example, in the mountain areas, this has also been turned towards tourism):[40]

> Here people are attached to the fact that the products are local, fresh and produced locally ... especially here in the mountains in the Basque country ... people are attached to territory. Restaurant owners won't want to buy from guys in Bordeaux, for example, it is not like that.
>
> (Interview: French trout producer, 2012)[41]

These local food supply chains have been of particular importance to producers who have sought to make explicit the environmental quality of farming practices in connection with territorial sustainability, building connections within local communities through social relations around the local food circuit (Maréchal and Spanu, 2010: 35):

> Local dimension is key. For our firm, it is essential…. As I said before, I am not just a producer. I go out, drink my coffee, do all the commercial side of the business. I do the deliveries. You have to be present on the terrain. People buy our products because it is us. I am not glorifying myself, but it is because I come from here and they know me and it is local. It is like that in the Basque country …
>
> (Interview: French trout producer, 2012)

Second, through their French and EU-wide collective associations producers have also engaged in political work within the EU arena, especially on the issue of what they perceive as unfair competition from imported non-EU fish products. In general terms, the problem as they define it is that whereas French producers are expected to respect very high EU environmental standards governing production practices in order to sell their products on European markets, EU trade policy permits the market entry of non-EU products grown under less stringent environmental regulations and hence commanding lower prices:

> A point over which we have often fought has been to have common rules for all fish sold in Europe.
>
> (Interview: Trout producers' representative, 2011)[42]

> The massive import of seafood from third countries is not a sustainable response to the needs of European consumers.
>
> (President of the Federation of European Aquaculture Producers [FEAP], cited in IUCN, 2011: 16)[43]

Around these problematisations of competition with non-EU imports (e.g. pangasius from Vietnam), we can therefore conclude that territorially based sustainability arguments have not only been used as a symbol to create value and hence compete at higher prices. They have also been mobilised as a weapon seeking to protect home markets from 'external' competition.

Regulation: Taking place alongside territorial political work on the frontier politics of interdependence and territory has been important political work at the interface of public and private regulation which has defined sustainability in trout food products sold in French supermarkets. This work has taken place on a French-wide scale. More specifically, Aquitaine producers have joined forces with other French producers, supermarkets and WWF to voluntarily self-regulate product specification in contracts between producers and supermarkets. This has both complemented public rules governing trout farm/environment interactions (for example in regard to rules for organic produce) and substituted for an absence of public regulation governing the sustainability of fish feeds.

Propulsion towards company work on FM and FO replacement in Aquitaine has only partially come about due to arguments advanced by scientists and e-NGOs in global fora in the early 2000s. Indeed, already by 1996, producers in Aquitaine had negotiated product specification criteria with the supermarket Carrefour, developing their brand 'filière qualité' for trout.[44] This product specification had included rules that feed ingredients would be GM-free and that marine ingredients would be sourced from sustainably managed feed fisheries. Additionally, by the start of the 2000s, in-house company policies had begun to introduce practices of substituting FM and FO in trout feed diets, hence reducing FIFO ratios:

> We know very well ourselves what the issues are. And I have proposed changes in the recipes because I am convinced – perhaps I have been influenced by the discourse of NGOs – but I have my own analysis of things. If you see that you need 6 kilos of caught fish to make 3 kilos of farmed fish, you don't need NGOs to tell you that there is a problem with that and that you need to go towards substitution. It is common sense. But this is specific to [our company]. We are small … we are friends who share the same values.
>
> (Interview: French organic feed company, 2012)

Producers too stressed during interview that work on substitution had been on-going for several years. Indeed, they argued, farmed trout could now be constructed as 'more sustainable' than wild fish when it came to FIFO ratios:

> The next challenge is to show that we can produce a kilo of farmed fish with a kilo of caught feedfish. Even if a wild fish will eat much more than a farmed fish in order to supply a kilo of eatable fish … a line caught seabass of one kilo will have eaten fourteen kilos of fish during its life because there is a five year cycle to reach one kilo and we can do that in two years and it will not only have eaten fish, but also cereals and so forth.
>
> (Interview: French trout producer, 2012)[45]

Building on these on-going practices, in 2010 the collective inter-professional association CIPA launched discussions with a group of French supermarkets and in alliance with WWF, consumer associations, scientists (from INRA) and public officials (from the fish farming division of the Ministry of Agriculture) to negotiate a single product specification for all trout farmers in France.

Up until then, individual supermarkets had developed their own product specifications in a logic of competitiveness. In a declining growth industry, this resulted in a lack of fluidity in the market, making it hard for producers to sell to multiple outlets (when their products did not exactly match product specifications), at times resulting in empty supermarket shelves void of trout products. Additionally, many supermarkets (with the exception of Carrefour) had not yet fully defined their policies on sustainability. Seizing an opportunity, CIPA called for a dialogue on these very questions. When this was finally set up in 2011, the sustainability of feeds was a central issue put on the table.

In these negotiations, a number of choices were made concerning feeds. First, a compromise was reached on the use of PAPs in fish feed diets, namely that their use would not be permitted. On interview, Aquitaine producers and feed companies had argued in favour of PAPs, making arguments equivalent to the ones made by Scottish salmon producers:

> We would take a huge step forward in reincorporating PAPs. It is a scandal not to do it. It is an ecological and economic luxury. A nonsense. We are burning these products in cemeteries and we are going off to look for fish meal made in Peru. It does not make any sense.
>
> (Interview: feed company, 2012)[46]

> We are walking on our heads. Mad cow disease, we regulated that. Now we have proteins which are very good for the fish, there are no problems, but no we can't [use them]. It is very good for the fish. For their flesh, for their growth.
>
> (Interview: trout company, 2012)

These positions notwithstanding, in their negotiations over the product specification, producers have compromised. Yet part of the compromise has been that this choice will not be expressly communicated to consumers as a praiseworthy decision:

> However we are not going to communicate this decision, we are not going to say "in the product specification there are no products from land animals", we won't do that because we don't think that it is a valid argument.
>
> (Interview: Trout producers' representative, 2012)[47]

A second choice has been not to set additional private rules on the provenance of raw materials which go beyond global standards and internationally agreed public policies already in place. Even if individual companies have in some cases developed in-house 'black lists' of suppliers from whom they will not buy their product, additional collective rating systems have not been developed:

> For the moment, we haven't placed any constraints and also we don't necessarily consider that this is … that this is necessary. We are working from the principle … well, we are working from the assumption that industrial fisheries [for fish meal and fish oil], for the majority of these stocks anyway, are well managed, the problem is that they are a finite resource.
>
> (Interview: Trout producers' representative, 2012)[48]

This choice therefore differs from the one made by Scottish private actors addressing this same issue. Indeed, it is connected to a broader difference of approach compared with Scottish salmon. Throughout the negotiations on French trout product specification, producers have sought to set private self-regulation criteria operating from within a 'logic of equality' as distinct from a 'logic of freedom to compete' (Smith, 2016).

This has been advanced on a number of fronts. First, the very motives behind the initiative were to ensure an equal playing field amongst trout farmers at a time of industrial decline (and to protect French products). Second, during the negotiations, producers disagreed with the WWF concerning the overall strategy. Initially WWF had wanted to set high product specifications for a premium product (as has happened for Scottish salmon). However, producers were against this. Third, social constructions of the sustainability of trout as a food product have been premised upon a social desire not to create a niche market for (richer) middle classes by selling 'premium' trout, but instead to create a standard sustainable product which can be sold at an affordable price. It is within this logic of equality too that we can understand the choice not to encourage a competitiveness strategy grounded in knowledge battles (Bush and Duijf, 2011) over rating systems, but to rely instead on global standards.

Knowledge: Important knowledge work has taken place supporting the market positioning of companies and collective private self-regulatory decisions. On the issue of FM/FO substitution, leading scientists in INRA have developed expertise on animal nutrition and fish feeds (Kaushik and Troel, 2010). Working within EU-funded projects on substitution (such as AQUAMAX), these scientists have shown that trout performs well compared to other species on reduced FIFO ratios.

Scientific discussions on refined approaches to substitution have also been at the centre of the work undertaken by the GIS-pisciculture initiatives, ('GIS pisciculteurs demain') working on the future of fish farming. For example, this work has shown that complete substitution of marine ingredients can have depressive effects on fish growth, whereas the use of substitution at different stages of the life cycle can have positive performative effects (Le Boucher 2012). This has led to new research, for example on micro-algae as potential new substitutes. R&D within the feed companies has also provided a wealth of knowledge on substitution and animal nutrition, as has the more

Table 6.2 Summary of political work over sustainability as a food governance problem for Aquitaine trout

Aquitaine trout			
Social constructions of sustainability	Institutionalisation of ecosystem approaches problematising ecology–society interdependencies and increasingly linked to sustainability of 'the homeland' have been complemented by social constructions of inter-relations between socio-ecosystems in non-contiguous places within a logic of equality		
Frontier politics of polity interdependencies	<u>Territory</u> No challenges to dominant EU political authority; Private rules define social meanings of Aquitaine's territories' sustainable trout as food (PGI Basque); Producers invest in work on local markets and local food supply chains; Territory as a value to compete for higher prices and as a weapon to protect home markets	<u>Regulation</u> Co-regulation governing sustainability marketing criteria on local environmental impacts; Policy capture by economic interests on sustainability marketing criteria on fish feeds in logic of equality; Conflicts over PAPs as replacements	<u>Knowledge</u> Scientific expertise and in-house company knowledge support choices on product specification; Inter-professional body mobilises its acquired knowledge as expertise to legitimise its authority to set rules on product specification

operationally-focused approach of the smaller organic feed company, which has tested its feed products in 'live experiments' working with a producer on this specific question.

A wealth of different forms of knowledge on substitution as sustainability has thus been generated. Whereas for a long time public and private rules have driven research questions (Medale, 2012), rather than the other way around, the more recent initiatives taken by CIPA to set product specifications can be analysed as an example of institutional positioning. In these debates, CIPA has self-represented as a holder of knowledge in order to legitimise its authority to set rules working in conjunction with supermarkets and WWF.[49]

Greek seabass/seabream and sustainability as a food governance problem

Overview of the changing sustainability narratives

Unlike both Scottish salmon and Aquitaine trout, a central element in Greek government of sustainability as a food governance problem for seabass and seabream has not been an increase in private self-regulation. Rather, a main theme in the Greek case has been consistent attempts to stress biophysical equivalences between natural and farmed fish within public rules.

The social construction of Greek seabass and seabream as a food product has been coherent with an adapted social meaning of sustainability initially depoliticising local ecology–economy interdependencies and later depoliticising wider global inter-ecosystem effects. Since the beginning, the marketing of seabass and seabream products has focused on the environmental attributes of farmed fish, stressing the biophysical equivalences between wild and farmed products. Unlike Scottish salmon and Aquitaine trout, which have been sold mainly in domestic markets, Greek seabass and seabream have mainly been exported.[50] For some companies, only 10 per cent of their product is sold in the Greek market. The rest of the seabass and seabream has been sold onto a generic commodity market in Europe (e.g. France, Spain, the Netherlands, Germany, the UK, Denmark and Switzerland) and elsewhere (e.g. the US, Ukraine, Russia and China), competing mainly on quality and price. In so doing, multiple contracts have been set up between producers, supermarkets, wholesalers and other retail outlets.

In general, there has been very little product diversification or differentiation through the use of eco-labels or brand names.[51] Rather, large companies have sold their fish whole or gutted seeking to render the product as 'natural' as possible compared with their wild fish counterparts:

> You make seabass and seabream and they look like this … many years ago now, before we even worked … [and] they keep selling the same way. That is the way the fish look. The only thing that you have to try to be, is very close to nature. Don't try to modify them …
>
> (Interview: Greek seabass/seabream producer, 2012)

Against this dominant trend, smaller companies have created niche markets (e.g. organic markets) in Northern European countries with labelled products. However, even though this difference is communicated in marketing strategies, on interview representatives of both large and small companies tended to play down the differences between labelled and non-labelled fish. Furthermore, whereas producers and feed companies have engaged in FM and FO substitution in feeds, thus implicitly recognising aquaculture's potential role in contributing to the decline of feed fisheries' stocks, producers nevertheless still tend to underline the role of aquaculture as compensating for the decline in fisheries:

> If you compare the impact of fishing and the impact of aquaculture, then you go definitely towards aquaculture. And it is good that we have developed it.
>
> (Interview: Greek seabass/seabream producer, 2012)

According to these kinds of arguments, not only is aquaculture compensating for the loss of fisheries, but fish farming is better as it produces a constant supply: 'tasty, healthy fish available every day, all year round' (Greek company slogan).

Overall, there appears to have been little change in the governing approach to marketing over time. In response to global debates on the use of marine ingredients in feeds, producers and feed companies have adapted their practices through developing self-regulatory, in-house policies which acknowledge the need for FM and FO substitution for the sustainability of fish products. Some of these policies have come about because producers sell a proportion of their fish to the supermarket (e.g. Carrefour in Greece) and must therefore comply with supermarket product specifications on feeds.[52] Yet there was no evidence of collective private standardisation practices which problematised sustainability equivalent to those found regulating the marketing of Scottish salmon or Aquitaine trout.

Unpacking interdependencies

To explain this depoliticisation of nature–society interdependencies in seabass and seabream as a food product, I analyse once again the frontier politics of interdependence and territory, regulation and knowledge.

Territory: Overall, I found very little evidence of political work to alter the frontiers of territorial interdependencies. With the exception of engagement by Greek producers within EU debates on the issue of the level playing field concerning imports of non-EU products (which I discuss below), in general it would appear that individual company strategies have been directed at closing down controversies on connections between seabass and seabream as a food product and territory.

First, there has been no collective attempt to create a Greek identity for the farmed seabass and seabream. In fact, on interview, the main category of territorial thought which emerged was 'the Mediterranean' (as was the case for farm/environment interactions). As argued by interviewees 'it is not in our culture to join forces' (Interview: Greek seabass and seabream producer, 2012). Even when faced with competition from Turkish products (which I discuss below), this has not resulted in political work constructing market-facing collective identities:

> Q: Do you see yourselves as having a collective identity, like Scottish salmon?
> No, unfortunately we don't do that, … a few years ago [Scottish salmon had a] really good lobby to protect against Norwegian salmon … they needed a really good lobby to do that, why don't we do the same for Turkish seabass and seabream coming in?
> (Interview: Greek seabass/seabream producer, 2012)

This absence of sustained political work to give market visibility to territory does not of course mean that Greek actors are not proud of their product:

> Our products are the best in Europe.
>
> (Interview: Greek public actor, 2012)

Rather, they have not organised collectively to explicitly market their territorial identity in the selling of their products. This is partly because their marketing strategies have been based on constructions of sustainability which seek to depoliticise as much as possible the differences between farmed fish and wild fish, focusing attention instead upon biophysical equivalences between wild and farmed products. On interview, producers stressed the naturalness of their fish and the uncertainties which the fish face in the open environment of the Mediterranean sea. They laid emphasis on the fact that for them aquaculture has to work in harmony with nature to be successful and produce 'a healthy Mediterranean diet'. A good example of this thinking was revealed in discussions over the differences between fish marketed as organic and those which are not:

> What is a "biological" fish? All our fish are biological ... [there is] lots of money in this and no difference in the fish.
>
> (Interview: Greek seabass/seabream producer, 2012)

When asked on interview over their choices to be organic or not, producers responded that going organic was not a production problem, but

> [an] advertising and supporting problem. Production is easy.
>
> (Interview: Greek seabass/seabream producer, 2012)

> Usually the conventional production was ok, so it was a very small step to make it organic actually.
>
> (Interview: Greek seabass/seabream producer, 2012)

Scientists we interviewed agreed:

> Organic aquaculture is conventional production ... organic means there are less fish in the cages ... not much difference in practice.
>
> (Interview: Greek scientist, 2012)

Attention to the constant naturalisation of the product was also evidenced in anecdotes about how producers have worked on creating new markets in other countries selling their fish to look as natural as possible:

> When we visited the market for Moscow, seabass and seabream were in supermarkets but treated in a terrible way. So we decided we had to do

something to increase sales there. We sent a guy there and went to all the supermarkets to 'train the market in Moscow'. No-one there to do it for us – we did it ourselves.

(Interview: Greek seabass/seabream producer, 2012)

This constant strategy has also played out in the creation of local markets selling fish with company brand names which have also focused on the naturalness of the product:

[We were] the first company to supply the market with the name of the company on the fish. This is quite dangerous because we are not sure quite what happens when you are delivering your products, how many days the other guy is going to have them in the fridge and all of that.

(Interview: Greek seabass/seabream producer, 2012)

Whereas in their marketing strategies, Greek producers have tended to depoliticise ecology–economy interdependencies, this has not been the case over the issue of the level playing field and competition from non-EU imported products. For Greek producers, this has particularly crystallised around issues of trade rules governing imports of Turkish products competing directly on Mediterranean markets:

Greek legislation is in accordance with the EU. It is strict where it should be and more strict than what is happening in the rest of the world so it automatically creates a disadvantage for European producers; it has [a] very environmental and animal welfare orientated focus. But it is not the same in Turkey, for example, which produces and imports and exports to the EU. And all the health standards and all the laws that control – the fish from the farm to the consumer is also strictly regulated. It is good for the consumer but it is bad because it is not the same for the imported products.

(Interview: Greek seabass/seabream producer, 2012)

It creates unfairness for Greek producers and producers [from] other countries. Because they don't have such strict rules to follow. They have much lighter rules, lower costs and they sell their product in the European Union also, without paying anything extra.

(Interview: Greek seabass/seabream producer, 2012)

However, even though Greek producers have engaged over this inter-territorial issue, both through their producers' federation and through FEAP, this action has not in turn led to political work to add value to 'Greek' products. Strategies of 'green consumerism' which directly make connections between the territorial location of production and sustainability of products were for the most part absent (unlike in the case of Scottish salmon and Aquitaine trout).

Regulation: In general, there has been very little political work to destabilise the interface of public and private regulation governing seabass and seabream as a sustainable food product. Not only were public rules governing farm/environment interactions considered sufficient, but also policy capture by economic interests on the issue of fish feeds has been absent. However, the lack of voluntary collective private self-regulation or standardisation has not meant that producers have not been aware of sustainability arguments on FM/FO substitution. On the contrary, they have been very aware of these debates and individual companies have developed in-house policies for substituting FM and FO in their feeds:

> We have to find ways to finish our production without pressuring the [feed fishery] stocks.
>
> (Interview: Greek seabass/seabream producer, 2012)

Greek producers have also engaged in debates on substitution through PAPs:

> We talk about sustainability. We don't allow animal by-products, which for carnivorous animals such as fish is exactly what they should be eating. We throw it away. We talk about animal welfare, but for carnivorous animals we make them eat soya and corn which is bad for their immune system. We have an increase in diseases in the Mediterranean, clearly because of the fact we have completely stressed the animals with their nutrition. We are not looking at the problem holistically. It is completely unsustainable.
>
> (Interview: Greek seabass/seabream producer, 2012)

On the question of feeds, large Greek companies have developed vertical integration company policies, establishing their own feed companies. Substitution has been an important strategy for these companies, reducing levels of FM and FO in diets. Yet this has not led to bespoke diets. On the contrary, feed companies have sought to establish a basic standard feed rather than engaging in a politics over multiple possible meanings of product quality:

> What we are doing in the last five years, is that we keep a basic feed standard and we try within this standard to find the best possible price.
>
> (Interview: Greek seabass/seabream producer, 2012)

Furthermore, Greek producers have not sought to work collectively on setting standards or rating systems to govern the sustainability of feed sources. Like Aquitaine trout producers, they rely on global standards governing global markets and claim not to buy from non-certified suppliers:

> The only non-certified sources for FM and FO are really dodgy, you wouldn't want to buy it anyway. The main suppliers of FM/FO are controlled by the FAO. So it is sustainable in this way.
>
> (Interview: Greek seabass/seabream producer, 2012)

This strategy has also been the case for sourcing vegetable ingredients like soya:

> Soya as well. Because we buy non-GM soya, there are very few producers in the world because of the pressure from the Americans, so they are definitely certified for everything existing nowadays, especially since soya has been the most scrutinised material for always.
>
> (Interview: Greek seabass/seabream producer, 2012)

Whereas larger companies have sold their product on a commodity market, smaller companies have engaged in different marketing strategies, acquiring organic and sustainability labels for their products and often selling into niche markets in Northern Europe. These companies have also tended to rely on standards set by other actors. Moreover, although they had adopted this 'green consumerism' strategy to differentiate themselves from the larger competitor companies, they revealed a certain suspicion over the accountability of the label:

> Private certifiers get paid from the label so what incentive to check and remove the label?
>
> (Interview: Greek seabass/seabream producer, 2012)

For these companies too, even if in their marketing strategies they appear to be establishing alternative constructions of sustainability, in fact there were many commonalities in the way both they and large companies have depoliticised the differences of a fish farmed under a label versus a fish not farmed under a label.

Table 6.3 Summary of political work over sustainability as a food governance problem for Greek seabass and seabream

Greek seabass and seabream			
Social constructions of sustainability	Initial depoliticisation of local ecology–economy interdependencies supplemented by depoliticisation of wider global inter-ecosystem effects		
Frontier politics of polity interdependencies	Territory	Regulation	Knowledge
	Politicisation of contents of EU trade policy but not seeking to re-assign authority; Territory not explicitly marketed in selling of product; Equivalences sought between wild and farmed fish	Public rules sufficient for sustainability marketing criteria; Conflicts over PAPs as replacements	Scientific expertise and in-house company knowledge to support marketing strategies

<u>Knowledge</u>: Of course, the generation of new knowledge on FM and FO substitution has been critical for implementing in-house policies. However, except for small companies operating in niche markets, it has not been explicitly used to support marketing strategies. As I have argued, these strategies have been primarily premised upon selling farmed fish which demonstrate environmental attributes based on biophysical equivalences with wild fish. To these ends, along with knowledge on feed performance, other types of knowledge generated to support these marketing policies have been in relation to consumer demand and prices.

Conclusions

In this chapter I have examined sustainability as a food governance problem for Scottish salmon, Aquitaine trout and Greek seabass and seabream. The literature on this question raises at least three potential issues: environmental attributes and wild and farmed fish equivalences; green consumerism; and product quality. The chapter has contributed to debates taking place within the respective literatures by providing a detailed analysis of how actors in each of the case study sites have actually framed the problem for themselves and set policy instruments to govern them.

I have identified similarities across territories and sectors over awareness of sustainability issues in the sourcing of raw materials for feeds. However, the research has also revealed core differences, notably over the marketing of the environmental attributes of farmed fish products. Indeed, whereas in all three cases the 'shadow ecologies' (Swanson, 2015) of fish farming aquaculture were recognised, this did not result in each case in their being governed through local private instruments acting on markets.

In Scottish salmon, sustainability as a food governance problem was initially conceptualised through narratives which depoliticise local ecology–economy interdependencies and with no marketing strategies communicating the environmental attributes of salmon products. However, combined attacks on salmon as food by scientists and e-NGOs destabilised these narratives. As we have seen, in response, a new coalition of actors came together and re-institutionalised sustainability in voluntary private self-regulation. This time actors gave social meaning to sustainability, recognising both local ecosystem impacts as well as those in distant places.

Changes in the market in line with this changed institution have been in the form of product differentiation. Shifts in narratives shaping sustainability and 'green consumerism' have been brought about through by joining them up with transforming polity narratives on territorial, regulation and knowledge interdependencies. This happened through political work defining the social meanings of 'Scottish' sustainable salmon within the framework of EU food safety policy (e.g. the PGI for Scottish salmon); co-regulation governing sustainability marketing criteria on local environmental impacts; and policy capture by economic interests around the sustainability marketing criteria for fish feeds. In addition, NGOs made political usage of their expertise on feeds to both disrupt the institution and then re-position themselves as experts on it. This further allowed them to gain access to discussions over product specification. In this political work, actors have constructed inter-relations between socio-ecosystems in non-contiguous places as a problem to be regulated. However, this has been carried out in a manner which supports a market logic of 'freedom to compete' amongst economic actors.

This stands in contrast to the trajectory of institutionalisation of sustainability as a food governance problem for the marketing of Aquitaine trout. Whereas the initial institutionalisation of ecosystem approaches problematising local ecology–society interdependencies has changed very little since the beginning, it has been increasingly linked to sustainability of 'the homeland'. This has happened especially through actor constructions of declining growth and the desire to better valorise their product as one possible response. Over time, this social construction of sustainability has been further supplemented by new social constructions of ecology–society interdependencies in far-flung places.

As was the case for Scottish salmon, the building of dominant narratives within the sustainability institution of Aquitaine trout has been accompanied by political work shifting territorial, regulation and knowledge interdependencies. Territory has been deployed by actors both as a value to compete for higher prices and as a weapon to protect home markets. In some cases this has resulted in institutionalisation through territorially explicit labels (e.g. PGI Basque trout). In particular, producers have invested in work on local markets and local food supply chains. Political work changing polity regularities in line with sustainability meanings has also included co-regulation to govern sustainability marketing criteria for local environmental impacts; policy capture by economic interests in relation to sustainability marketing criteria for fish; and the generation of scientific expertise and in-house company knowledge to support choices on product specification. However, in comparison with Scottish salmon, these have been institutionalised within a logic of equality between fish farming producers.

Finally, in regard to Greek seabass and seabream, the initial depoliticisation of local ecology–economy interdependencies has been complemented by narratives depoliticising wider global inter-ecosystem effects. This has resulted in little product differentiation and the seeking of equivalences between wild and farmed fish in marketing strategies. These depoliticisations have been maintained through actor work politicising the contents of EU trade policy but not seeking to re-assign authority; legitimising public rules as sufficient for sustainability marketing criteria and the generation of knowledge about prices to support marketing policies.

Consequently, some general conclusions comparing European aquaculture policy on the governance of farmed fish as a sustainability food problem can be drawn. In Scotland, aquaculture policy on this issue has been made up of a mixture of public and private regulation. For example, EU public rules on food and feed safety along with Scottish public rules on the environmental and visual impacts of farms have all been complemented by a raft of private instruments and standards set within private labels, supermarket brands and private partnerships. Unlike in the governance of both fish farm/environment interactions and access to fish farm sites, governing spaces created to discuss the contents of these policies are not only public Scottish ones. Rather, they also include partnerships between economic actors and some e-NGOs, and extend across the industry to include feed companies and retailers.

In Aquitaine, where aquaculture policy on food is also governed by a mixture of public and private regulation, we observe by contrast that private regulation has been used by economic actors to compensate for the perceived shortcomings in public policies and is driven by notions of equality in the market place. In addition, although governing spaces opened up through private self-regulation are French-wide ones (which also include retailers and WWF), there has been a specific local quality to this private regulation aimed at the co-construction of both local markets and territory.

In Greece, aquaculture policy on food has mainly been limited to public general rules governing both feeds and environmental impacts. There has been little evidence of collective private self-regulation. Even when tensions have accumulated over competition with non-EU imports, these have not been translated into self-regulatory practices on product differentiation. This means that whereas in both Scotland and Aquitaine private self-regulation has opened up space for debating the sustainability of farmed fish as food amongst economic and civil society actors, this has not been the case in Greece to date.

Notes

1 In the language of Jullien and Smith (2014), Chapters 4 and 5 are concerned with institutional and political work within and across the 'Production' and 'Employment' institutions of the industry, whereas in this chapter we move to consideration of institutional and political work within and across 'Commercial' and 'Production' institutions (see the third section of Chapter 3).
2 Except for those companies which grow fish for re-stocking and restoration purposes.
3 This problem has been discussed by Swanson as the problem of 'shadow ecologies' (2015).
4 See also Carter *et al.*, 2014. The politics of eco-labelling of aquaculture products has been further studied, for example, in regard to failed attempts to create a label valorising the positive biodiversity effects of carp production in the Dombes region in eastern France (Wezel *et al.*, 2013)) and in the identification of contractual relationships between producers and supermarkets as critical ones to facilitate product differentiation and product quality in shellfisheries (Charles and Paquotte, 2009).
5 These concepts are closely linked. Research by Young *et al.* (1999) and Fernández-Polanco and Luna (2010) has shown that food safety and food sustainability are closely associated concepts in consumers' minds affecting their purchasing choices when buying farmed fish.

6 They were banned in 2001.

7 Although companies sell to a range of markets worldwide (e.g. UK, US, France, Canada, Chile, China, Dubai, Japan) the majority of Scottish salmon is sold domestically in the UK market (interviews): 80 per cent of the fish bought by consumers in the UK is bought in the supermarket (Ernst & Young 2008: 80).

8 Salmon product quality under the LR label is predominantly judged concerning its taste, texture and flavour (interviews).

9 One commentator, David Millar, has presented the resultant response by the Scottish industry as one of 'spin', reading into the political work of actors a collective conspiracy against 'science as truth' (see: www.academia.edu/2939514/Spinning_farmed_salmon). Otherwise, the leading scientists of the study were reported at the time as saying: 'We think it's important for people who eat salmon to know that farmed salmon have higher levels of toxins than wild salmon from the open ocean' (Professor Ronald Hites). 'My choice would be, if I were to seek out farm-raised Atlantic salmon, to select north or south American sources, based on these data' (Professor Barbara Knuth; co-author). These scientists were both quoted in an article entitled 'Scottish farmed salmon "is full of cancer toxins"' by Roger Highfield, Science Editor of *The Telegraph* newspaper (9 January 2004). Available at: www.telegraph.co.uk/news/uknews/science/science-news/3317547/Scottish-farmed-salmon-is-full-of-cancer-toxins.html.

10 At the time of writing, three feed companies supply Scottish salmon producers: EWOS, Biomar and Skretting.

11 This is in contrast to institutional work on Aquitaine trout which I consider to be structured more by a 'logic of equality' (Smith, 2016) – see below.

12 Whereas EU food safety regulation covers the contents of fish feeds on grounds of safety, it does not regulate the provenance of fish feed ingredients on grounds of sustainability (Carter, 2015a).

13 Although feeds were potentially covered in the 2009 Scottish strategy within the working group on marketing and image, in fact this group has ultimately left this issue to the market place. Not all e-NGOs are in agreement with this decision and there have been some tough exchanges on this very topic.

14 Within the Scottish sectoral strategy on aquaculture – see Chapters 4 and 5.

15 They have also engaged over a level playing field for products in the EU market place. For example, in the European Commission consultation process (2007–2009) leading to the development of a European strategy for the sustainable development of aquaculture, the SSPO (Scottish Salmon Producers Organisation) argued against what it considered to be unfair competition resulting from the lack of a level playing field between EU and non-EU products in European markets. This has been a central issue raised within the FEAP (Federation of European Aquaculture Producers) on sustainability and competitiveness in EU trade policy:

> There are obviously marketing pressures. There is the question mark of importing products and services from abroad when they don't have the same standards as operating in Europe, even though they are supposed to. So for example, in Norway they can use six or even eight chemicals for treating sea lice and we have two or three. And they have other medicines as well. There are disparities in the legislation which cause problems because ultimately it means that they can produce their fish more efficiently and cheaper and sell them onto the European market place cheaper than we can.
>
> (Interview: Scottish public actor, 2011)

16 This is also the general view of US and European Fat Processors and Renderers Associations who produce PAPs, as well as the European Commission – see Carter, 2015b.

17 Scottish actors have engaged in multiple ways: through FEAP; in external consultation processes organised by the European Commission; and in seminars on feeds organised by the European Parliament (see Carter and Cazals, 2014).

18 Equivalent to the scheme in place for wild fisheries which is the Marine Stewardship Certification (MSC).

19 This included the launch of a campaign advocating the ASC to UK supermarkets and encouraging them in the future to buy only ASC-certified salmon (interviews).

20 For example, in the tilapia dialogue, farming companies from Ecuador and Central America were more present than those from Asian countries, 'making it impossible for the group to create rules that could encompass global production, as was the initial ambition' (Havice and Ilas 2015: 33).

21 For example, one supermarket sells 'responsibly sourced' salmon. There is no label on the package, but the supermarket message to consumers guarantees that any salmon they buy, whether fresh or in a can, is responsibly sourced (interviews).

22 For example, the Soil Association organic label requires *inter alia* lighter stocking densities of fish and limited use of chemical treatments for sea lice.

23 Or, to put it another way, around both a politics of substitution and a politics of provenance (Carter, 2015a).

24 To illustrate: one can imagine that feeds need to contain different ratios of ingredients to meet different needs. For example, feeds need to contain a minimum amount of FO to guarantee that the product provides the consumer with adequate levels of Omega 3 and 6 so that it can be sold on human health grounds; they also need to have reduced FM and FO ingredients to be sold on sustainability grounds; and at the same time, fish feed contents must be nutritionally valuable for the fish to maintain fish health requirements (including animal welfare) (interviews).

25 In response to Hites *et al.*, 2004, feed companies have reduced the sourcing of their marine ingredients from fisheries in the NE Atlantic.

26 For example, international bodies like the Marine Ingredients Organisation (IFFO) have set their own standards on responsible sourcing of FM and FO from feed fisheries to sell on the global market.

27 Products have been sold with different labels including Organic AB and Label Rouge.

28 98–99 per cent of French trout producers belong to the inter-professional association, CIPA (interviews).

29 80 per cent volume of trout as food goes to the retailer.

30 French trout producers were indirectly affected by the publication of the Hites *et al.* article. However, this was managed internally and has not been the main challenge affecting sustainability institutionalisation practices.

31 Aquitaine trout's main market is the domestic French market.

32 Competing with salmon steaks and smoked salmon at lower prices (interviews).

33 Original French: 'On s'est séparé de l'univers nordique et on est maintenant sur l'univers franco-français.'

34 Original French: 'C'est une marque ou on décline véritablement le développement durable, l'aspect originaire territoriale truite française mais plus loin truite des Pyrénées, truite de Pays Basque, truite d'Aquitaine en déclinant tous les aspects santés, Oméga 3 car la truite est moins gras que le saumon et naturellement riche en Oméga 3. Et on fait un dernier point important, accent important, c'est en achetant l'OVIVE vous pérennisez de l'emploi local.'

35 Unlike in Scotland, however, it is not the direct action of e-NGOs on the French trout market which has placed this issue on the agenda: the organic feed company Le Goussant has been particularly instrumental in influencing these debates in Aquitaine.

36 Although the French government did issue a public statement when the EU re-authorised PAPs claiming that this EU authorisation notwithstanding, French farmed fish would not be fed PAPs. Indeed, the French government had voted against the authorisation. In an article in the newspaper *Le Monde* the then President of France, François Hollande, was quoted as saying to a meeting on agriculture that France did not want PAPs, insisting instead on the right to 'continue to feed fish in a traditional manner' (original French: 'continuer à nourrir les poissons de manière traditionnelle'). Available at: www.lemonde.fr/planete/article/2013/02/23/plats-cuisines-hollande-souhaite-un-etiquetage-obligatoire_1837703_3244.html.

37 For example, Aqualande's product brand name OVIVE has territorial designations such as OVIVE 'Truite Fumée Pyrénées' and OVIVE 'Truite Fumée Aquitaine'.

38 Original French: 'On va justifier la qualité de notre produit en montrant qu'on est bien intégré dans le milieu et démontrer que notre activité n'est pas délocalisable. On va construire de la valeur autour de ça.'

39 Original French: 'Je crois qu'il y a des changements récents pour favoriser la consommation locale. Il y a des patrimoines locaux qui reviennent.'

40 This has also involved product diversification, e.g. one company has produced high-end luxury processed food products (e.g. curried trout rillettes) which are sold to local shops.

41 Original French:

> Ici les gens sont attachés au fait que les produits soit locales, frais et produits localement … surtout ici dans les montagnes du Pays Basque … les gens sont attachés au territoire. Les restaurateurs ne vont pas acheter des gens de Bordeaux, par exemple, c'est pas ça.

42 Original French: 'Un point sur lequel on bat beaucoup, c'est d'avoir des règles communes pour tous les poissons vendues en Europe.'

43 Original French: 'L'importation massive de produits aquatiques de pays tiers n'est pas une réponse durable aux besoins des consommateurs européens.'

44 Producers have also set rules for Label Rouge labelling of their products (fresh and smoked). Yet, in the same spirit as their Scottish counterparts, French actors have more recently criticised these Label Rouge specifications on grounds of sustainability: 'Label Rouge is a product that doesn't have legs. It goes against the spirit of sustainability.' (Original French: 'Label Rouge, c'est un produit mort né. Il va à l'encontre de la durabilité.') (Interview: French feed company, 2012).

45 Original French: 'Le prochain défi serait d'arriver à montrer qu'avec un kilo de poisson pêché en mer on arrive à produire un kilo de produit aquacole. Même si un poisson sauvage consomme bien plus qu'un poisson d'aquaculture pour arriver à fournir un kg de poisson comestible … un bar de ligne d'un kilo aura mangé quatorze kilos de poissons durant sa vie parce que le cycle pour atteindre un kilo c'est à cinq ans et nous on l'aura fait en deux ans et il n'aura pas mangé que des poissons mais aussi des céréales etc.'

46 Original French: 'On franchirait un pas énorme en réincorporant les PATs [Protéines Animales Transformées]. C'est un scandale de ne pas le faire. C'est une montreuse écologique et économique. Un non-sens. On brûle ces produits dans des cimetières et on va chercher des farines faites au Pérou. C'est un non-sens.'

47 Original French: 'Mais par contre on va pas communiquer dessus, on va pas dire "dans le cahier des charges unique il n'y a pas de produits d'animaux terrestres", on va pas en faire état parce qu'on considère que c'est pas un argument valable.'

48 Original French: 'Pour le moment on n'a pas mis de contraintes, et puis on ne juge pas forcément que ce soit … que ce soit nécessaire. On part du principe … enfin, on part du constat que la pêche minotière, quand même pour la plupart des stocks, est bien gérée, le problème c'est que c'est une ressource finie.'

204 *Institutionalising sustainability*

49 Unlike in Scotland where e-NGOs have mobilised knowledge to assign authority to themselves over sustainability and feeds, WWF France has not developed this same strategy. Part of the reason has been that they hold very little expertise on trout farming as such (interviews).
50 Seabass and seabream is the second largest export sector of Greek agriculture, after olives and olive oil (Barazi-Yeroulanos, 2010).
51 For example, labelled organic production is less than 1 per cent of total Greek seabass and seabream production (Polymeros *et al.*, 2014).
52 For example, there is the Carrefour brand 'filière qualité' for seabream and seabass (interviews). Initially this demanded use of feed from international feed companies, but more recently feed produced by Greek feed companies is accepted as equivalent.

References

Barazi-Yeroulanos, L. 2010. *Regional Synthesis of the Mediterranean Marine Finfish Aquaculture Sector and Development of a Strategy for Marketing and Promotion of Mediterranean Aquaculture (MedAqua-Market)*, Studies and Reviews: General Fisheries Commission for the Mediterranean No. 88. Rome: Food and Agriculture Organisation (FAO).
Bendiksen, E., Johnsen, C., Olsen, H., Jobling, M. 2011. 'Sustainable aquafeeds: progress towards reduced reliance upon marine ingredients in diets for farmed Atlantic salmon (*Salmo salar L.*)' *Aquaculture*, 314: 132–139.
Borraz, O. 2007. 'Governing standards: the rise of standardization: processes in France and in the EU' *Governance*, 20(1): 57–84.
Bureau, D. 2006. 'Rendered products in fish aquaculture feeds'. In Meeker, D. (ed.) *Essential Rendering: All about the Animal By-products Industry*, Arlington: The National Renderers Association; The Fats and Proteins Research Foundation; The Animal Protein Producers Industry, pp. 180–194.
Bush, S., Duijf, M. 2011. 'Searching for (un)sustainabilty in pangasius aquaculture: a political economy of quality in European retail', *Geoforum*, 42: 185–196.
Byelashov, O., Griffin, M. 2014. 'Fish in, fish out: perception of sustainability and contribution to public health', *Fisheries*, 39(11): 531–535.
Carter, C. 2012. 'Integrating sustainable development in the European government of industry: sea fisheries and aquaculture compared'. In Shuibhne, N. and Gormley, L. (eds) *From Single Market to Economic Union*, Oxford: Oxford University Press, pp. 289–314.
Carter, C. 2015a. Who governs Europe? Public versus private regulation of sustainability of fish feeds', *Journal of European Integration*, 37(3): 335–352.
Carter, C. 2015b. 'De-politicizing Europe: collective private action and sustainable Europe'. In Carter, C. and Lawn, M. (eds) *Governing Europe's Spaces: European Union Re-Imagined*, Manchester: Manchester University Press, pp. 172–194.
Carter, C., Cazals, C. 2014. 'The EU's government of aquaculture: completeness unwanted'. In Jullien, B. and Smith, A. (eds) *The EU's Government of Industries: Markets, Institutions and Politics*, Abingdon: Routledge, pp. 84–114.
Carter, C., Ramírez, S., Smith, A. 2014. 'Trade policy: all pervasive but to what end?' In Jullien, B. and Smith, A. (eds) *The EU's Government of Industries: Markets, Institutions and Politics*, Abingdon: Routledge, pp. 216–240.
Costa-Pierce, B. 2010. 'Sustainable ecological aquaculture systems: the need for a new social contract for aquaculture development', *Marine Technology Society Journal*, 44(3): 88–112.

Crawford, M. 2010. 'The importance of long chain omega-3 fatty acids for human health'. Paper presentation at IG-CCBSD conference 'Can A Growing Aquaculture Industry Continue To Use Fishmeal And Fish Oil In Feeds And Remain Sustainable?', European Parliament, Brussels, 3 March.

Drakeford, B., Pascoe, S. 2008. 'Substitutability of fishmeal and fish oil in diets for salmon and trout: a meta-analysis', *Aquaculture Economics & Management*, 12(3): 155–175.

Ernst & Young. 2008. *Etude des performances économiques et de la compétitivité de l'aquaculture de l'Union européenne*, Report for DG MARE, December.

Fernández-Polanco, J., Luna, L. 2010. 'Analysis of perceptions of quality of wild and cultured seabream in Spain', *Aquaculture Economics & Management*, 14(1): 43–62.

Fernández-Polanco, J., Luna, L. 2012. 'Factors affecting consumers' beliefs about aquaculture', *Aquaculture Economics & Management*, 16: 22–39.

Gillund, F., Myhr, A.I. 2010. 'Perspectives on salmon feed: a deliberative: assessment of several alternative feed resources', *Journal of Agricultural and Environmental Ethics*, 23: 527–550.

Greenpeace. 2005. *A Recipe for Disaster: Supermarkets' Insatiable Appetite For Seafood*, London: Greenpeace.

Havice, E., Ilas, A. 2015. 'Shaping the aquaculture sustainability assemblage: revealing the rule-making behind the rules', *Geoforum*, 58: 27–37.

Hites, R., Foran, J., Carpenter, D., Hamilton, M., Knuth, B., Schwager, S. 2004. 'Global assessment of organic contaminants in farmed salmon', *Science*, 303: 226–229.

Huntington, T.C. 2004. *Feeding the Fish: Sustainable Fish Feed and Scottish Aquaculture*, Report to the Joint Marine Programme Scottish Wildlife Trust and WWF Scotland and RSPB Scotland.

IUCN. 2011. *Guide pour le développement durable de l'aquaculture: Réflexions et recommandations pour la pisciculture de truites*. Gland, Switzerland and Paris, France: IUCN.

Jackson, A. 2007. 'Challenges and opportunities for the fishmeal and fishoil industry', *Feed Technology Update*: 2.

Jaffry, S., Pickering, H., Ghulam, Y., Whitmarsh, D., Wattage, P. 2004. 'Consumer choices for quality and sustainability labelled seafood products in the UK', *Food Policy*, 29: 215–228.

Jullien, B., Smith, A. (eds) 2014. *The EU's Government of Industries: Markets, Institutions and Politics*, Abingdon: Routledge.

Kaushik, S.J., Troell, M. 2010. 'Taking the fish-in fish-out ratio a step further', *Aquaculture Europe*, 35(1): 15–17.

Le Boucher, R. 2012. 'Adapter les poissons aux aliments végétaux: programme VEGEAQUA', Paper presented at the Conference 'Towards competitive and sustainable fish farming', organised by INRA, CIPA, Itavi and Sysaaf, 27 February, Paris.

Mansfield, B. 2003. 'From catfish to organic fish: making distinctions about nature as cultural economic practice', *Geoforum*, 34: 329–342.

Maréchal, G., Spanu, A. 2010. 'Les circuits courts favorisent-ils l'adoption de pratiques agricoles plus respectueuses de l'environnement?', *Le courrier de l'environnement de l'INRA*, 59: 33–45.

Mauriac, L. 2013. 'Farines pour poissons et viande de cheval, même logique pour Batho', *Rue 89* [online]. Available at: www.nouvelobs.com/rue89/rue89-plan ete/20130217.RUE3340/farines-pour-poissons-et-viande-de-cheval-meme-logique-pour-batho.html.

Natale, F., Hofherr, J., Fiore, G., Virtanen, J. 2013. 'Interactions between aquaculture and fisheries', *Marine Policy*, 38: 205–213.

Naves, P. 2016. *Du court, du local! Une sociologie du gouvernement de la filière fruits et légumes*, PhD Thesis, University of Bordeaux, 2 December.

Naylor, R.L., Goldburg, R.J., Primavera, J.H., Kautsky, N., Beveridge, M.C.M., Clay, J., Folke, C., Lubchenco, J., Mooney, H., Troell, M. 2000. 'Effect of aquaculture on world fish supplies', *Nature*, 405 (6790): 1017–1024.

Naylor, R., Hardy, R., Bureau, D., Chiua, A., Elliott, M., Farrelle, A., Forstere, I., Gatlin, D., Goldburg, R., Huac, K., Nichols, P. 2009. 'Feeding aquaculture in an era of finite resources', *Proceedings of the National Academy of Sciences (PNAS)*, 106(36): 15103-15110.

Olesen, I., Alfnes, F., Bensze Røra, M., Kolstad, K. 2010. 'Eliciting consumers' willingness to pay for organic and welfare-labelled salmon in a non-hypothetical choice experiment', *Livestock Science*, 27: 218–226.

Polymeros, K., Kaimakoudi, E., Mitsoura, A., Nikouli, E., Mente, E. 2014. 'The determinants of consumption for organic aquaculture products: evidence from Greece', *Aquaculture Economics & Management*, 18(1): 45–59.

Smith, A. 2016. *The Politics of Economic Activity*, Oxford: Oxford University Press.

Swanson, H. 2015. 'Shadow ecologies of conservation: co-production of salmon landscapes in Hokkaido, Japan, and southern Chile', *Geoforum*, 61: 101–110.

Tacon, A., Metian, M. 2008. 'Global overview on the use of fish meal and fish oil in industrially compounded aquafeeds: trends and future prospects', *Aquaculture*, 285: 146–158.

Tacon, A.G.J., Metian, M. 2009. 'Fishing for feed or fishing for food: increasing global competition for small pelagic forage fish', *AMBIO: A Journal of the Human Environment*, 38: 294–302.

Tusche, K., Wuertz, S., Susenbeth, A., Schulz, C. 2011. 'Feeding fish according to organic aquaculture guidelines EC 710/2009: influence of potato protein concentrates containing various glycoalkaloid levels on health status and growth performance of rainbow trout (*Oncorhynchus mykiss*)', *Aquaculture*, 319: 122–131.

Wezel, A., Chazoule, C., Vallod, D. 2013. 'Using biodiversity to valorise local food products: the case of fish ponds in a cultural landscape, their biodiversity, and carp production', *Aquaculture International*, 21: 1395–1408.

Whitmarsh, P., Wattage, P. 2006. 'Public attitudes towards the environmental impact of salmon aquaculture in Scotland', *European Environment*, 16: 108–121.

Young, J., Brugere, C., Muir, J. 1999. 'Green grow the fishes-oh? Environmental attributes in marketing aquaculture products', *Aquaculture Economics & Management*, 3(1): 7–17.

7 Conclusions

The 'tangled politics' of sustainability interdependence

Recent scientific rumours that 'sustainability' is outmoded as a concept require a considered response. Certainly, in the case of fish farming aquaculture, sustainability is far from over as a policy debate. As I have both argued and demonstrated throughout this book, in this industry, sustainability is a core political struggle. This has been evidenced in actor conflicts over the governing of a range of issues, including over the environmental impacts of fish farms; interactions between fish farming production and biophysical processes; environmental landscape aesthetics; nature as a societal benefit; coastal development; the environmental attributes of farmed fish products; and 'shadow ecologies' between production and sourcing socio-ecosystems (Chapters 4, 5 and 6).

Moreover, even though there are specificities in the case of aquaculture's relationship to sustainability which I have also explained, I anticipate that in the politics of other nature-based industries, such as fisheries and forestry, sustainability remains a central policy challenge. Additionally, debates in Europe on the implementation of sustainability are likely to become ever more critical for other economic sectors interacting with oceans and coasts as the full implementation of the EU Marine Strategy Framework Directive gets underway. Already in New Aquitaine in France the setting of actions and indicators to give effect to the objectives of this directive are raising challenging 'sustainability' questions for a range of industries, including gravel extraction, shipping and coastal tourism.[1] Rather than being obsolete as a political issue, on the contrary, 'sustainability' is likely to remain a pressing societal concern.

That being said, the main line of argumentation advanced in this book has been to agree that if 'sustainability' is not dead as an idea, nonetheless our way of thinking about it has to change. More precisely, I have argued that sustainability requires both updating as a concept and recomposing through its being grounded in a broader theory of political change.

In response, I have first treated 'sustainability' heuristically as an institution (public action principle) and de-centred it. This was achieved through carrying out an extensive literature review on sustainability. Following from this review, I identified four 'ideal-type' narratives on nature–society interdependencies which were likely to be mobilised by actors to render it meaningful for them. These were: (i) sustainability as depoliticised ecology–economy interdependencies; (ii) sustainability as politicised ecology–economy–society interdependencies; (iii) sustainability as ecosystem approaches problematising dynamic ecology–society interdependencies; and (iv) sustainability as de-growth re-ordering ecology–society–economy hierarchies.

However, my argument did not stop there. Drawing on political sociological approaches analysing political change, I went further to argue (and demonstrate) that sustainability's contents have not just been shaped through political struggles over social constructions of changing nature–society interdependencies. In connection with these, as I have shown, actors have also mobilised around a frontier politics at the heart of broader polity interdependencies between territories, between public and private regulation and between knowledge forms. This has resulted in sustainability choices – over fish farm/environment interactions, over access to fish farm sites, and over farmed fish as a food product – becoming *entangled* within broader institutionalisations of territorial power, states, markets and civil society, science and knowledge. In other words, through re-conceptualising sustainability from within public policy analysis which stresses policy–polity dialectics, I have highlighted how interdependence and territory, interdependence and regulation and interdependence and knowledge are all endogenous factors in the reproduction of sustainability politics.[2]

In this way, I have reframed the research object as 'sustainability interdependence'. In doing so, I have sought to make a clear statement that studying interdependencies at the heart of sustainability politics does not mean limiting the line of inquiry to those between nature and society. Rather, it means expanding the inquiry to show how actors put narratives on nature–society interdependencies into relation with other political interdependencies potentially operating at different temporal and spatial scales. The politics of sustainability interdependence is therefore not merely over the policy substance of sustainability. Crucially, it is also about the political authority to influence choices in the first place and maintain or disrupt domination when governing. As the empirical chapters reveal, through studying what I can now call a *'tangled politics'* of sustainability interdependence, research can confirm that this process is never complete but has become part and parcel of what actors governing fish farming aquaculture do.

Following from these general remarks, in the rest of these conclusions I first revisit sustainability interdependence as an analytical framework. I then briefly compare the main elements of sustainability interdependence approaches governing Scottish salmon, Aquitaine trout and Greek seabass and seabream. I finish with some general remarks on how thinking in terms of sustainability interdependence potentially opens new ways of debating aquaculture's place in the economy and society. In particular, I argue that the application of this approach encourages reconsiderations of the 'ecosystem approach' to aquaculture in the light of case study findings.

Sustainability interdependence as an analytical framework

A central claim advanced in this book, and demonstrated in the empirical chapters, is that studying how actors in fish farming aquaculture have governed the central problems they face through the prism of sustainability interdependence allows research to grasp the complexity of causality in politics. This is because how actors institutionalise all four types of interdependence simultaneously – between nature and society; between territories; between public and private regulation; and between knowledge forms – determines how the specific problems with which they are faced are defined by them, governed and legitimised. This creates a 'tangled' politics which needs unpacking.

My entry points for unpacking this politics have been the political sociological concepts of 'institution' and 'institutional and political work'. These have allowed me to study sustainability as being socially constructed by actors in response to different problems facing their industry. Using the concepts of political and institutional work, I have shown how sustainability has been created, maintained, disrupted and/or re-built through interactions between actors setting instruments over long periods of time.

In the three empirical chapters of this book, political work observed revealed a contingent and antagonistic politics conforming with definitions of politics as proffered by Mouffe (2000, 2005), Kauppi et al. (2016) and Itçaina et al. (2016). The periodic arrival at unity over definitions of problems and contents of instruments did not mean that a fully inclusive rational consensus between actors had been achieved. Certain actors and their arguments dominated governing choices over time and were more or less tolerated by other actors and arguments. Some actors and their arguments have remained 'outside' from the start (such as radical e-NGOs), but their action has nonetheless not been without effect. This has meant that in all cases, those governing did so in a situation of tension and alternatives whereby political work was undertaken by those dominating to protect the status quo as much as it was by those seeking to destabilise it. Whilst some actors struggled to open up problems and instruments to contestation, others sought to close them down.

Moreover, as I have also argued, the politics of sustainability interdependence studied from this perspective took place within institutionalised relationships. Indeed, the findings reveal strong interactions between actors and institutions. In respect of each case studied – whether fish farm/environment interactions, access to fish farm sites or sustainability as a food governance problem – actors were not passive in the face of anonymous social processes or laws. They 'worked' on them and through them, using a myriad of resources and identities to legitimise their arguments.[3]

The application of the analytical framework of sustainability interdependence (as detailed in Table 3.1) enabled me to capture political work both at the meso level and as it has accumulated over time. This encouraged the production of synthetic accounts of changing policy trajectories and their causes. Indeed, at this meso level, the framework proved to be robust and heuristic in guiding the disentanglement of the politics studied. This in turn enabled me to tease out synthetic comparative causal stories on the governing of problems, through the coding and sorting of my findings.

Inevitably, there have been limitations in this process. Whereas operating at a meso level of analysis permits aggregation and comparison, and the making of connections between actors and institutions across time and space, in aggregating my data I have had to synthesise from quite complex and detailed debates, and in so doing, I have had to let go of some of the content. Additionally, the actors interviewed were often quite aware and reflexive about the tangled nature of the interdependence politics in which they were engaged, including its myriad contradictions.[4] Actually this was often talked about through wry humour: the interdependent world in which they engaged at times amused them as much as it frustrated them. In the end, I found it hard to present this content in the case studies. I have sought to compensate by using anonymous actor quotations as much as I could to not only present actors' arguments, but also to present how they 'lived' and experienced the governing of sustainability interdependence.

In the next section I return to my empirical findings having re-organised them this time around sectors and territories, as distinct from policy problems.

Sustainability interdependence governing Scottish salmon, Aquitaine trout and Greek seabass and seabream

As I have reconfirmed, my approach to studying sustainability governing has been to apply sustainability interdependence as a causal and comparative framework analysing actors' political and institutional work around specific policy problems (see Chapters 4, 5 and 6). In this section of the conclusions, I wish to take a further analytical step through drawing some more general conclusions on how policy choices around specific problems are altering broader policy–polity relations in each territory and sector. This is to conclude on the governing *outcomes* of sustainability choices made in Scottish salmon, Aquitaine trout and Greek seabass and seabream.

In the elaboration of the framework in Chapter 3, I included a set of criteria which would allow a comparison of outcomes of policy choices (Table 3.1). These are: (i) policy substance and tensions; (ii) assignment of political authority; (iii) tensions between scales; and (iv) forms of democracy. I will now revisit each territory and sector in turn to conclude on each of these elements. This is to come back again to the question I set at the start of the book on how Scottish salmon farming, Aquitaine trout farming and Greek seabass and seabream farming are both structured by and structuring of politics. Of course, I have answered this question in the empirical chapters in respect of specific policy problems. In these conclusions, I have chosen to reconstitute

Table 7.1 Comparative contingent sustainability interdependencies of Scottish salmon, Aquitaine trout and Greek seabass and seabream

	Policy substance and tensions	Assignment of political authority	Tensions between scales	Forms of democracy
Scottish salmon	Anthropocentric ecosystem approaches: tensions around utilitarian versus distributional politics; Conflicts over hierarchies of nature's societal benefits (salmon farming versus alternative coastal economies)	SG and producers dominate, but in coalitions with other stakeholders	Attempts at synergies between scales: all problems brought to the scale of Scotland	Tensions: Participatory versus technical democracy
Aquitaine trout	Anthropocentric ecosystem approaches challenged by quasi-ecocentric approach to water governance; Conflicts over hierarchies of nature's societal benefits (trout farming versus wild, rural rivers)	French government (Ministry of Ecology) and coalitions dominate, but producers have important regional partnerships	Mismatch of scales	Tensions: Participatory democracy for water governance versus classical lobbying for trout farming
Greek seabass/seabream	De-politicised ecology–economy narratives in tension with anthropocentric ecosystem approaches; Conflicts over hierarchies of nature's societal benefits (seabass/seabream farming versus tourism)	Greek state ministries and producers dominate, but attacked by local communities and courts	Mismatch of scales	Tensions: Anarchic non-compliance and public contestation versus participatory democracy

my narratives to come at this question again from a different angle comparing territories and sectors overall. This is to ask what are the main outcomes of actors' political work co-producing the fish farming sector and society in each case? I stress that I am not seeking in any way to present here essentialist images of static 'models' of comparative territories and sectors. There is not one model of Scottish salmon to be compared with one model of Aquitaine trout or Greek seabass and seabream. On the contrary, institutions change, are contingent, and require studying over time. As we have seen, there are both spatial and temporal scales of sustainability interdependence to be recognised. However, I consider that it is useful to offer a generalised 'country' comparison by way of conclusion, thereby potentially opening up new lines of inquiry for future studies on European fish farming.

Governing sustainability in Scottish salmon

As we have seen in the case studies carried out in the empirical chapters (and as I have summarised in the country-specific tables in the Annex), sustainability in Scottish salmon has changed its social meaning over time. Initially given dominant meaning through narratives depoliticising ecology–economy interdependencies, by the end of the 2000s, these narratives no longer carried authority. They had been replaced by anthropocentric ecosystem approaches problematising ecology–society interdependencies.

However, tensions have persisted between actors and especially over choices and usage of policy instruments. For example, some producers primarily discuss aquaculture's relationship to nature in terms of ecology–economy interdependencies. Some e-NGOs (and local actors) espouse ecocentric ecosystem approaches. Some policy instruments are more directed at technological forcing,[5] some are about Environmental Strategic Assessment[6] and others still are about balancing ecosystem services.[7] Additionally, anthropocentric ecosystem approaches are being implemented alongside a Scottish Government commitment to growth, which has also (as we have seen) created tensions at the heart of the sustainability institution. This strong growth policy has been further coupled with extensive private self-regulation within a 'logic of freedom to compete' (Smith, 2016). So although we can conclude that the dominant sustainability policy governing Scottish salmon espouses an anthropocentric ecosystem approach, we must also qualify that within this policy there are strong tensions between a utilitarian versus a distributive politics.

Political work over policy substance has additionally confirmed who governs. The governing of Scottish salmon has taken place following a critical re-assignment of authority to govern from the UK and the EU to Scottish public and collective private actors. Both in respect of governing a range of issues pertaining to fish farm/environment interactions and access to farm sites (environmental policy, fish health, biodiversity, planning), both public and collective private actors in Scotland have mobilised to re-assign authority to Scotland. Further, even over fish feeds, where public actors have not (yet) made any moves to govern through public rules, collective private actors in Scotland have nonetheless sought to prioritise Scottish approaches.

Drawing on the case studies, the Scottish Government (including Marine Scotland) and producers have emerged as dominant actors. Yet clearly, a range of other actors have had a critical role shaping the contents of the sustainability institution. This is because even though the Scottish Government and producers are consistently dominant actors, different coalitions of actors govern different problems. For example, the setting up of the working groups under the sectoral strategy for aquaculture brought e-NGOs and salmon fishery boards inside governing processes. Even if these groups do not perceive themselves as dominant therein, nonetheless their presence and their arguments has contributed to the re-shaping of sustainability politics. Additionally, through the planning mechanism, regulatory agencies and local authorities have been critical actors. Finally, in regard to product specification for the selling of salmon, it has been UK supermarkets, feed companies and some e-NGOs working with producers which have influenced choices.

It follows from this that Scottish devolution has mattered in critical ways for the governing of the sustainability of salmon farming. This has been enjoined by clear attempts made by actors to find synergies between different types of governance: administrative, sectoral, public/private and the socio-ecological system. For all problems, irrespective of their type, it has been attempted to govern them at the scale of Scotland.

Finally, political work has resulted in important shifts towards the democratisation of Scottish salmon. Key issues on environmental impacts and planning have been taken out of the hands of private estate owners and brought into the public space. Further, governing choices over on-going issues facing the industry – such as escapes of fish, sea lice diseases or new planning rules – have been allocated to participatory governing working groups for negotiation. However, on these questions we observe a tension between participatory forms of democracy and technical ones. On the one hand, the scientification of controversies, especially by dominant actors, has been commonplace. On the other, the extensive voluntary private self-regulation undertaken by private collective actors (at times working with e-NGOs) on salmon product specification has captured key elements of policy by economic interests. When this happens, the role of citizens becomes one of consumers (Stoker, 2009: 85) – and the role of e-NGOs as representatives of the 'general interest' of Scotland can be called into question (Chartier and Ollitrault, 2005).

In summary, these conclusions over the institutionalisation of sustainability interdependence of Scottish salmon can be compared with others examining transformations in post-devolution Scotland over public policy design (Cairney, 2006), and the government of other industries (Carter and Smith, 2008, 2009). In all these cases, Scottish actors have sought to push the limits of devolution's frontiers to define 'Scottish' approaches, irrespective of the formal frontiers of the legal settlement (MacPhail, 2008).

Additionally, tensions I observed at the heart of the anthropocentric EA narrative over the importance of Scottish society (and local communities) versus empowerment of economic actors, on the one hand, and participatory democracy versus neoliberal governance on the other, have also been identified (and further critiqued) (Law and Mooney, 2006). For example, Law and Mooney have argued that the new Scotland, whilst appealing to norms of equality, is also highly structured around values of individual entrepreneurship (2006). Indeed, Scottish Government support for other sectors by easing planning mechanisms for coastal development has been documented (Warren and McFadyen, 2010).

Moreover, this mixing of 'older traditions of political sovereignty and self-definition' with inclusive and open government has been even stronger since the SNP took up office in Scotland in 2007 (Ozga and Dubois-Shaik, 2015: 117). This therefore confirms key moments of change in Scottish salmon's causal story, particularly concerning access to farm sites, when (new) policies favouring the growth of salmon farming were put in tension with pre-existing ones favouring the benefits of nature to develop 'diverse coastal economies'. Governing the future sustainability of Scottish salmon will likely continue to centre upon these tensions.

Governing sustainability in Aquitaine trout

By contrast with Scottish salmon, and as we have seen from the case studies (see also the tables in the Annex), Aquitaine trout was initially governed through anthropocentric ecosystem approaches regulating water quality. However, and as I have explained, the implementation of the Water Framework Directive in France occasioned a powerful coalition of actors mooting quasi-ecocentric EA narratives for governing water and creating ambiguities over trout farming's future role in the future of the rivers. Consequently, in regard to key issues of farm/environment interactions, access to farm sites and nature as multiple benefits, tensions over these different conceptions of interdependencies between trout farming and rivers created conflicts and uncertainty.

These tensions over whether production can be included in the image of the rural river or not must be understood in a situation of industrial decline of trout farming and its lack of growth in rural areas. Debates over trout farming's future in line with anthropocentric ecosystem approaches have been kept alive through the political work of trout producers – forming alliances with others, as I discuss below – setting self-regulation and generating new knowledge as a counterpoint to dominant public rules on farm/environment interactions. Producers have also sought to codify their conception of sustainability in product specification, capturing policy on sustainability marketing criteria. This has resulted in their keeping alive, both in their arguments and in private instruments, their interpretation of sustainability which has been premised on a strong attachment to the 'homeland' within a 'logic of equality' (Smith, 2016).[8]

In Aquitaine trout, the dominant actors governing the sustainability of trout farming have been the officials within the French government's Ministry of Ecology and other actors within the environmental water governance framework, including actors within river basin committees, ONEMA and water agencies – at times in loose alliances with recreational fishers and organisations protecting rivers and water. However, producers have mobilised collectively to create shadow coalitions within Aquitaine (and France), joining forces in a number of ways: with regional public actors (over fish health and environmental impacts); with officials in the fish farming division of the Ministry of Agriculture; with the international NGO IUCN debating obstacles to the growth of their industry; and finally also with WWF and French supermarkets setting product specification.

In contrast to the governing of Scottish salmon, there has been a mismatch of scales whereby the decentralisation of water governance does not overlap with the governing parameters of Aquitaine trout, which remain institutionalised on a French-wide scale. Further, there has been no real carving out of political space to debate trout farming's future – for example, through the development of a sectoral strategy along the lines of the Scottish one. Trout farming continues to be governed mainly through transversal and general regulation and consequently through the latter's dominant social constructions of nature–society interdependencies, which are often at odds with those of trout producers.

The democratic consequences of being governed through a distant top-down state approach has been ambiguity for producers and their shadow networks over whether they are merely 'victims' of choices made by others, or whether they can self-assign ownership of their future in dialogue with others. More critically still, it is hard to conclude that trout farming has been democratised. Thus far democratic mechanisms of the public inquiry around EIAs (Blatrix, 2009) have not served to facilitate a broader societal discussion on farm site authorisations (if this was even their purpose). Often, the public inquiry has been captured instead by the interests of powerful, rich, recreational fishers, who have otherwise dominated discourses seeking to turn 'valleys into sanctuaries' (Germaine and Barraud, 2013). Similarly, even if a participatory democracy of water governance within river basin committees has been observed, these organisations address broader inter-sectoral issues along the water body wherein trout farmers are in a clear minority. Consequently, Aquitaine trout stakeholder participation takes place either through classical French centre–periphery relations or within an 'integrated' process at the scale of the watershed or water plan.

In summary, the governing of Aquitaine trout, whereby a distant top-down state still dominates, turns out not to be specific either to this territory or this sector in France. Indeed, many contradictions have been observed by others concerning French decentralisation. On the one hand, territorialisation and deliberation processes under the environmental programme Grenelle have been shown to have fragile impacts (Whiteside *et al.*, 2010). Whilst decentralising, state actors have nonetheless continued to focus political work so as to maintain their leadership whilst governing at a distance and developing new forms of state governance of local territories (Poupeau, 2013). Using the same words which were used to describe state officials' strategies for governing rivers' futures, Poupeau has described state strategy in energy policy as being one of the 'reconquest' of French institutional space.[9]

Meanwhile, even within water governance more generally, scholars have argued that not many restoration projects have been based on participatory democracy as envisaged by the WFD (Germaine and Barraud, 2013). This has caused them to argue further that ecological continuity objectives still require to be linked to territorial projects (Germaine and Barraud, 2013). Indeed, the dominance of regulative science in the implementation of the WFD has been experienced by trout producers as equivalent to 'national' [*sic*] narratives premised upon an 'abstract image of uniformity' rejecting 'local specificities', as documented by Pasquier in relation to decentralisation politics in other domains (2015: 182).

Governing sustainability in Greek seabass and seabream

Greek seabass and seabream was governed from the start through actor narratives depoliticising ecology–economy interdependencies. These narratives are still very important ones today, upheld in arguments advanced both by some producers and some Greek public actors. However, following attempts at destabilising this 'win–win' philosophy through attacks led by local communities 'outside' the governing institution, these narratives have been gradually joined by ecosystem approaches problematising ecology–society interdependencies. This has resulted in tensions between these two narratives at the heart of the institution. However, unlike in the Scottish case where attacks on the win–win philosophy were mediated by bringing the attacking actors within the governing framework, in the case of Greek seabass and seabream, local communities remain mostly on the 'outside'.

Instead, change has been brought about through strategies developed by some Greek public officials who, drawing on EU policy norms, have introduced new integrated ecosystem policy tools to govern conflicts over Greek seabass and seabream. This has resulted in the setting up of public consultation processes to depoliticise attacks on the sustainability institution. The mobilisation of these policy tools notwithstanding, conflicts persist amongst actors over which economic activities coastal nature should be benefitting – fish farming aquaculture or coastal tourism? In the Greek shift towards anthropocentric ecosystem approaches, debates on the hierarchy of the societal benefits from nature therefore remain rife and ecosystem spatial planning has not resolved conflicts over choices.

The dominant actors governing Greek seabass and seabream have been central state public actors and producers. However, these two groups often act independently of one another. This has been compounded by a lack of coordination amongst the multiple different ministries involved and which has been a feature of Greek executive politics more generally (Featherstone and Papadimitriou, 2013). Moreover, there has been no development of a Greek sectoral strategy similar to the Scottish one to establish organisational structures for actor discussions over the industry's future. Producers have engaged in dialogue with government officials through classical lobbying, personal contacts and anarchical behaviour.

In this situation, actor strategies of protest have been effective, whereby through community mobilisation and judicial review of the legality of licences, as we have seen, administrative court power has come into conflict with executive power (and won). Otherwise, shadow actors include supermarkets in those countries to which seabass and seabream is exported. These actors hold market power under a depoliticised sustainability marketing policy selling seabass and seabream onto a commodity market. This has resulted in the use of multiple contracts and commodity purchases, thus encouraging competition on price and quality and power for supermarkets to switch suppliers with ease.

In Greece, as in Aquitaine, there has been a mismatch of scales of governance, whereby nationally determined plans are attempting to overlap simultaneously onto local socio-ecosystems and water bodies and onto the fish farming sector, both of which operate at different scales. This has created confusion for many actors over whether the new policy tools of the ecosystem approach are integrated ones, or whether they are rather designed to mitigate the fall-out over the weakness of depoliticised ecology–economy approaches. In other words, whereas the central state still dominates public action, it continues to have little handle upon economic and societal actors.

Finally, in Greece, as I have shown in the empirical chapters, sustainability politics for a very long time took place in a broader situation of legislative anarchy, which was already documented by others examining the environmental regulation of seabass and seabream farming nearly two decades ago (Conan, 2000). In general, actors have talked about a past situation of non-compliance with the rules, corruption and non-implementation of legislation. Other problems have been a lack of continuity between fast changing governments and the absence of a strong environmental movement in Greece (Fousekis and Lekakis, 1997), which could have introduced norms of participatory democracy into debates over sustainability (Chartier and Ollitrault, 2005).

However, this situation is changing and in Greek seabass and seabream its democratisation can be said to have commenced with recent participatory democratic mechanisms introduced to govern access to farm sites, revealing a separation of the autonomy of the state from civil society, as is part and parcel of state administrative reform more widely (Papadakis, 2012). As argued by Spanou (2008), Greek state reform has been proceeding through trial and error, where a central challenge has been, and continues to be, addressing the mistrust prevalent in relations between citizens and the administration. Modernisation reforms underway are opening up past 'clientelist practices and networks' (Featherstone and Papadimitriou, 2013) and addressing segmented government, therefore potentially offering actors new resources and identities to bring about change in democratising Greek seabass and seabream.

Opening up new ways of debating aquaculture and sustainability

Following on from these conclusions, in this final section of the book I would like to propose different ways in which the analysis conducted here can open up new ways of debating both aquaculture and sustainability. A consistent finding of this book is that the relationship of fish farming to sustainability is neither spontaneous nor natural, but is governed. In this way, I have made the case for treating the governing of this industry as meriting analytical attention in its own right – something which has hitherto been absent in an otherwise multidisciplinary literature on fish farming aquaculture. Throughout, my aim has been to problematise aquaculture's sustainability politics, rather than just assuming that its politics is something which is well known (Kauppi *et al.*, 2016).

This has led me to emphasise that all account of the effectiveness of regulation governing this industry will always be hampered if they either confuse institutions with formal rules or stay at a descriptive level of the rules' contents without taking into account how they have been produced. The findings also underline that, if centred solely upon public policy regulation, explanations of aquaculture will remain incomplete given the wealth of voluntary private self-regulation managing production and commercial practices.

The example of Environmental Impact Assessments (EIAs) is useful here. As we have seen, the use of EIAs as key regulatory instruments has been documented in all three case studies, yet with radically different applications. In particular, there were important differences in whether EIAs were regulating a relationship strictly between the economic activity and its local ecological impact or whether they were being applied in a broader context shaping societal debates about fish farming's place within a broader socio-ecosystem. In short, any focus on regulation or policy without attention to its politics will likely generate only partial conclusions, thus hampering our understanding of this industry's relationship to sustainability.

Throughout, the book has sought to steer clear of making any normative judgement about the value of fish farming aquaculture per se. In this manner, my approach to reveal a politics of aquaculture in this book contrasts with those accounts of the global consequences of the establishment of this industry and which appear to reject it outright (e.g. Clausen and Clark, 2005; Longo, 2012).

The starting point of these latter authors' work is that, as a biotechnological industry, fish farming aquaculture is automatically creating what they term a 'metabolic rift' between humans and the environment, disconnecting people from their historical attachment to nature (Longo, 2012). In my case studies, any negative societal and/or ecological consequences of aquaculture were not universally inevitable but depended upon how they were governed.

In this regard, the results presented in the book's chapters reveal a more complex set of findings about how actors hold differently ' "anthropized"/ artificial nature ... within a technological society' (Choné *et al.*, 2017: 2). Analysing sustainability from the critical perspective of sustainability interdependence thus enables research to go beyond a global critique of aquaculture as a societal failure to more nuanced understandings (White, *et al.*, 2017) of the tangled politics of its political economy and the democratic consequences of governing choices.

As I have shown in respect of each public policy problem governed (fish farm/environment interactions; access to farm sites; farmed fish as food), the contents of European aquaculture policies are composed of comparable, contested objectives: grow the industry (Scottish salmon); control the industry (Aquitaine trout); contain the industry (Greek seabass and seabream). These policies are shaped through multiple mechanisms and instruments of political intervention operating at different administrative, sectoral, transectoral and socio-ecological scales. They also operate on both public and private domains of governance and are sustained through mobilisation of a variety of territorial identities and knowledge types.

As I have summarised above, in Scotland, there is strong regional governance in both public and private domains; in Aquitaine trout, French-wide public governance is in tension with regional private regulation; whilst in Greece, Greek state and regional public regulation have been in tension and confronted by judicial activism. In Scotland, public and private regulatory interdependencies operate from within a logic of freedom to compete; in Aquitaine, these interdependencies operate from within a logic of equality; and in Greece, there is strong public regulation and very limited private regulation. Nonetheless, there is also a logic of freedom to compete.

In all three sectors and territories, regulatory science dominates policy choices, although this is increasingly being supplemented (and in some cases challenged) by industry and NGO expertise (both of which are also frequently pitched against one another). In Scotland, aquaculture policy has been shaped through tensions between participatory versus technical democracy; in Aquitaine it has been shaped through tensions between participatory democracy for water governance versus classical lobbying for trout farming; in Greece it has been shaped through tensions between anarchic non-compliance and public contestation versus participatory democracy.

Viewing sustainable aquaculture policy in this light, the approach adopted in this book can contribute to on-going 'thick' analyses of sustainability (Adger *et al.*, 2003) treating sustainability as a potential vector of power which builds identities and political authority (Van Tatenhove 2013, 2016). As I have shown, these are not limited to ecology, economy and society interdependencies and identities, but extend to territorial ones, the public versus the private and attachments to different kinds of science and knowledge. Thinking this way in terms of sustainability interdependence thus takes us beyond binary 'light green' versus 'dark green' or 'weak' versus 'strong' categorisations of sustainability. In their place, it offers a categorisation based upon differentiation, controlling for shifts between different types of narratives about nature–society interdependencies.

As the different case studies have revealed, these shifts can be explained by the way in which actors have navigated the organisation of power through exploiting the frontier politics of territorial, regulatory and knowledge interdependencies. For example, through actor navigation of the formal organisation of territorial jurisdiction to redistribute territorial resources of political authority; through authority capture by economic interests and civil society working in the market place; and through political and scientific struggles to promote the 'value' of particular kinds of science and knowledge for decision-making. Ultimately, therefore, the original analytical model developed and applied in the book enables us to differentiate sustainability interdependence narratives in aquaculture around specific criteria.

In this context, an important and unexpected finding of the book is that whereas there are clear differences between territories and sectors, there has nonetheless been a *convergence* towards institutionalising *ecosystem approaches to aquaculture*. Indeed, the extent of change revealed in each case study site has meant that in each territory and sector, actors have globally moved beyond the first two narratives of sustainability and today are operating from within ecosystem narratives which problematise dynamic ecosystem ecology–society interdependencies. Of course, this is not to imply that this narrative has replaced all others – this has not been a linear process of change. On the contrary, different actors and instruments continue to give expression to other meanings of sustainability. However, the point of settlement or 'unity in diversity' in respect of each policy problem examined has come about around this third ecosystem ecology–society narrative.

Any discussion on European aquaculture must therefore grapple with the political nuances of ecosystem approaches (EAs). This is particularly pertinent because no universal 'ecosystem approach' could be identified: rather, the EAs applied were quite different in respect of each territory and sector. Indeed, different forms of EA were explained through the way in which the different interdependencies of territory, regulation and knowledge had been navigated. Consequently, departing from common assumptions that ecosystem approaches reduce conflicts through encouraging rational deliberation over aquaculture's future, instead I found that the politics implementing these approaches was antagonist. Indeed, political tensions at the heart of this sustainability narrative were as important as those between this narrative and the others.[10]

On the one hand, in Aquitaine trout, frictions over quasi-ecocentric versus anthropocentric interpretations have dominated the politics of the governing of this industry. On the other hand, oppositions *within* anthropocentric ecosystem approaches were also in evidence, depending on whether actors were conceptualising and regulating these from within utilitarian or distributional theories of politics. These oppositions played out especially within the politics of Scottish salmon, but could also be observed as a difference between the politics of private self-regulation of Scottish salmon versus those of Aquitaine trout. Tensions were also revealed by actors over hierarchies of nature's benefits in all three territories and sectors, and in particular over nature as benefiting production processes versus nature as benefitting tourism or recreation.

For all these reasons, the application of the analytical grid opens new ways of debating ecosystem approaches applied to aquaculture and invites a line of questioning on its underlying politics. In this way, the book therefore ultimately encourages a critical shift in how we appraise European fish farming. Rather than staying with general examinations of the governing of European fish farming per se, it encourages continued detailed analysis of the *ecosystem politics* of fish farming. This is to further encourage a reframing of conventional questions on this industry.

Whereas hitherto two types of question have dominated debates – 'do we want fish farming?' (e.g. Longo, 2012); and 'have we applied ecosystem approaches to aquaculture?' (e.g. Soto *et al.*, 2008) – the findings presented in this book beg the asking of different questions. These are: 'who decides on whether or not to expand fish farming and how to regulate it?'; 'where are the political spaces of authority governing fish farming and who inhabits them/who is left out?'; and 'which form of ecosystem politics (utilitarian/distributional) of fish farming is being advocated and with what consequences?'.

In the future, sustainability interdependence could therefore be usefully applied as a framework to further unpack the politics of governance through ecosystem approaches in more detail – also by extending its application to other industries. Put differently, the analysis of the debate on aquaculture's relationship to sustainability begun here now needs testing by applying it to see how similar controversies play out in other sectors and polities.

This brings me to a final set of comments about the heuristic value of the model of sustainability interdependence which potentially extends beyond aquaculture towards a more general analysis of sustainable adaptation to global change. As the application of the model makes clear, political work on interdependencies is a central feature of adaptation. With its analytical focus on the tensions inherent in this political work on interdependence, the application of the framework therefore provides a way forward for grasping controversies of adaptation – often missing in systemic approaches (Olsson *et al.*, 2015).

Operating at a meso level of analysis, its scientific foundations can also contribute to comparative research controlling for a range of variables explaining how newly recast ecological challenges are rendered governable in diverse settings. In this way, model findings can be translated into targeted policy advice in situations of diversity (i.e. moving away from 'one size fits all' solutions). As we have seen, the application of the model can also produce results on the democratic dimension of adaptation and whether adaptation is brought about through technical, participatory or representative democracy.

The model also offers potential for new synergies across disciplines. Applied to fish farming aquaculture in this book, the analytical framework has already proven its capacity to capture the tangled politics of sustainability interdependence around different industrial challenges. This matters, because, as we have seen throughout, fish farming aquaculture is an industry which crystallises current conundrums on innovative solutions to global change through biotechnological innovation which, on the one hand, solve one set of ecological problems, whereas on the other, they can generate new controversies.

This same conundrum can be found in other cases, e.g. in regard to hydro-electricity, which is both a solution for green energy and a major cause of biodiversity loss through river fragmentation. Although political analyses of these challenges can produce new and hitherto absent data, and hence make an important contribution to our understanding, ultimately, these are not challenges which can be comprehensively addressed by one discipline acting alone or in isolation from others. This is particularly the case when the resolution of the challenge turns on the governing of interaction of ecosystem services and their hierarchy.

For these reasons as well, the approach developed in this book potentially offers a novel approach to conducting interdisciplinary research beyond political science and in this way opens up new ways of debating aquaculture.

Notes

1 As scholars have recently argued, tourism poses its own challenges to climate change and has its own environmental impact. These issues have hitherto remained largely under-problematised in societal debates (Dissart *et al.*, 2015; Hall *et al.*, 2013).
2 Or in other words, they are not just the background political context.
3 As I demonstrate throughout, this does not mean that they always 'won' their arguments.
4 Of course actors did not use the word 'tangled'!
5 First narrative: sustainability as depoliticised ecology–economy interdependencies.
6 Second narrative: sustainability as politicised ecology–economy–society interdependencies.
7 Third narrative: sustainability as ecosystem approaches problematising dynamic ecology–society interdependencies.
8 Compared with the Scottish logic of 'freedom to compete'.
9 Original French: 'Une stratégie de maintien voire de reconquête de sa place dans l'espace institutionnel français' (Poupeau, 2013: 469).
10 This concurs with recent work by Waylen *et al.* (2015) on the implementation of conservation management ecosystem approaches.

References

Adger, W., Brown, K., Fairbrass, J., Jordan, A., Paavola, J., Rosendo, S., Seyfang, G. 2003. 'Governance for sustainability: towards a "thick" analysis of environmental decision making', *Environment and Planning*, A(35): 1095–1110.

Blatrix, C. 2009. 'La démocratie participative en représentation', *Sociétés Contemporaines* 2(74): 97–119.

Cairney, P. 2006. 'Venue shift following devolution: when reserved meets devolved in Scotland', *Regional & Federal Studies*, 16(4): 429–445.

Carter, C., Smith, A. 2008. 'Revitalizing public policy approaches to the EU: "territorial institutionalism", fisheries and wine', *Journal of European Public Policy*, 15(2): 263–281.

Carter, C., Smith, A. 2009. 'What has Scottish devolution changed? Sectors, territory and polity-building', *British Politics*, 4(3): 315–340.

Chartier, D., Ollitrault, S. 2005. 'Les ONG d'environnement dans un système international en mutation: des objets non identifies?' In Aubertin, C. (ed.) *Représenter la nature? ONG et biodiversité*, Paris: IRD Editions, pp. 21–58.

Choné, A., Hajek, I., Hamman, P. 2017. 'Introduction: rethinking the idea of nature'. In Choné, A., Hajek, I., Hamman, P. (eds) *Rethinking Nature: Challenging Disciplinary Boundaries*, London: Routledge, pp. 1–10.

Clausen, R., Clark, B. 2005. 'The metabolic rift and marine ecology an analysis of the ocean crisis within capitalist production', *Organization & Environment*, 18(4): 422–444.

Conan, N. 2000. 'L'environnement dans la réglementation et la gestion des piscicultures en Grèce'. In Petit, J. (ed.) *Environnement et aquaculture, Tome II Aspects juridiques et réglementaires*, Paris: INRA, pp. 266–274.

Dissart, J-C., Dehez, J., Marsat, J-B. (eds). 2015 *Tourism, Recreation and Regional Development*, Farnham: Ashgate.

Featherstone, K., Papadimitriou, D. 2013. 'Core executive politics in Greece: the paradox of absent centralization'. Paper Presented at the 63rd Political Studies Association Annual International Conference, Cardiff, 25–27 March.

Fousekis, P., Lekakis, J. 1997. 'Greece's institutional response to sustainable development', *Environmental Politics*, 6(1): 131–152.

Germaine, M-A., Barraud, R. 2013. 'Les rivières de l'ouest de la France sont-elles seulement des infrastructures naturelles? Les modèles de gestion à l'épreuve de la directive-cadre sur l'eau', *Natures Sciences Sociétés*, 4(21): 373–384.

Hall, C., Scott, D., Gössling, S. 2013. 'The primacy of climate change for sustainable international tourism', *Sustainable Development*, 21: 112–121.

Itçaina, X., Roger, A., Smith, A. 2016. *Varietals of Capitalism: A Political Economy of the Changing Wine Industry*, Ithaca and London: Cornell University Press.

Kauppi, N., Palonen, K., Wiesner, C. 2016. 'Controversy in the garden of concepts: rethinking the "politicisation" of the EU'. Joint Working Paper Series of Mainz Papers on International and European Politics (MPIEP) No. 11 and Jean Monnet Centre of Excellence 'EU in Global Dialogue' (CEDI) Working Paper Series No. 3. Mainz: Chair of International Relations, Johannes Gutenberg University.

Law, A., Mooney, G. 2006. ' "We've never had it so good": The "problem" of the working class in devolved Scotland', *Critical Social Policy*, 26(3): 523–542.

Longo, S. 2012. 'Mediterranean rift: socio-ecological transformations in the Sicilian bluefin tuna fishery', *Critical Sociology*, 38(3): 417–436.

MacPhail, E. 2008. 'Changing EU governance: a new opportunity for the Scottish Executive?', *Regional & Federal Studies*, 18(1): 19–35.

Mouffe, C. 2000. *Deliberative Democracy or Agonistic Pluralism*. Wien: (Reihe Politikwissenschaft/Institut für höhere Studien, Abt. Politikwissenschaft 72. Available at: http://nbnresolving.de/urn:nbn:de:0168-ssoar-246548.

Mouffe, C. 2005. *On the Political*, Routledge: London.

Olsson, L., Jerneck, A., Thoren, H., Persson, J., O'Byrne, D. 2015. 'Why resilience is unappealing to social science: theoretical and empirical investigations of the scientific use of resilience', *Science Advances*, 1: 1–11.

Ozga, J., Dubois-Sahik, F. 2015. 'Referencing Europe: usages of Europe in national identity projects'. In Carter, C. and Lawn, M. (eds) *Governing Europe's Spaces: European Union Re-Imagined*, Manchester: Manchester University Press, pp. 111–129.

Papadakis, N. 2012. 'Réforme des administrations publique et locale: le cas de la Grèce', *Outre-terre*, 32: 36–369.

Pasquier, R. 2015. *French Regional Governance and Power in France: The Dynamics of Political Space*, Basingstoke: Palgrave Macmillan.

Poupeau, F-M., 2013. 'Quand l'état territorialise la politique énergétique: l'expérience des schémas régionaux du climat, de l'aire et de l'énergie', *Revue Politiques et Management Public*, 30(4): 443–472.

Smith, A. 2016. *The Politics of Economic Activity*, Oxford: Oxford University Press.

Soto, D., Aguilar-Manjarrez, J., Hishamunda, N. (eds). 2008 *Building an Ecosystem Approach to Aquaculture*, FAO/Universitat de les Illes Balears Experts Workshop. 7–11 May 2007, Palma de Mallorca, Spain. FAO Fisheries and Aquaculture Proceedings No. 14, Rome: Food and Agriculture Organisation (FAO).

Spanou, C. 2008. 'State reform in Greece: responding to old and new challenges', *International Journal of Public Sector Management*, 21(2): 150–173.

Stoker, G. 2009. 'What's wrong with our political culture and what, if anything, can we do to improve it? Reflections on Colin Hay's "Why We Hate Politics"', *British Politics*, 4: 83–91.

Van Tatenhove, J. 2013. 'How to turn the tide: developing legitimate marine governance arrangements at the level of the regional seas', *Ocean & Coastal Management*, 71: 296–304.

Van Tatenhove, J. 2016. 'The environmental state at sea', *Environmental Politics*, 25(1): 160–179.

Warren, C., McFadyen, M. 2010. 'Does community ownership affect public attitudes to wind energy? A case study from south-west Scotland', *Land Use Policy*, 27: 204–213.

Waylen, K., Blackstock, K., Holstead, K. 2015. 'How does legacy create sticking points for environmental management? Insights from challenges to implementation of the ecosystem approach', *Ecology and Society*, 20(2): 21.

White, D., Gareau, B., Rudy, A. 2017. 'Ecosocialisms, past, present and future: from the metabolic rift to a reconstructive, dynamic and hybrid Ecosocialism', *Capitalism Nature Socialism*, 12(2): 1–19.

Whiteside, K., Boy, D., Bourg, D. 2010. 'France's "grenelle de l'environnement": openings and closures in ecological democracy', *Environmental Politics*, 19(3): 449–467.

Annex

Summary tables of comparative sustainability
interdependencies of Scottish salmon, Aquitaine trout
and Greek seabass and seabream

Table A1.1 Comparative sustainability interdependencies of Scottish salmon

Fish farm/environment interactions			
Social constructions of sustainability	From de-politicised ecology–economy interdependencies to anthropocentric ecosystem approaches problematising dynamic ecology–society interdependencies		
Frontier politics of polity interdependencies	<u>Territory</u> Separation of Scottish approach and Scottish societal ownership; New identities and jurisdiction; Multi-positioning strategies on UK, EU scales	<u>Regulation</u> From private guidance on EIAs to public control; New public sectoral policy complemented by new private regulation	<u>Knowledge</u> Claims of knowledge gaps to de-stabilise initial institutionalisation; Scientification of controversies in re-institutionalisation; Conflicts over interactions between farmed and wild salmon

Access to fish farm sites			
Social constructions of sustainability	De-politicised ecology–economy interdependencies are replaced with anthropocentric ecosystem approaches. However, tensions emerge over hierarchy of ecosystem services being institutionalised		
Frontier politics of polity interdependencies	<u>Territory</u> Separation of Scottish approach; Political work by producers and SG to reduce conflicts over how different actors have connected visions of territorial futures with representations of the environment	<u>Regulation</u> From private planning to public control; Tensions over regulation as supporting nature and providing multiple benefits to society versus supporting nature and providing benefits to private economic actors	<u>Knowledge</u> Scientification of controversies in re-institutionalisation

Sustainability as a food governance problem			
Social constructions of sustainability	Attacks on sustainability as de-politicised local ecology–economy interdependencies lead to a re-institutionalisation implementing ecosystem approaches which recognise inter-relations between socio-ecosystems in non-contiguous places within a logic of freedom to compete		
Frontier politics of polity interdependencies	<u>Territory</u> No separation of Scottish public approach; EU public rules on food safety dominate; Private rules define social meanings of 'Scottish' sustainable salmon as food and private actors engage in multi-positioning strategies	<u>Regulation</u> Co-regulation governing sustainability marketing criteria on local environmental impacts; Self-regulation of sustainability criteria for fish feeds (policy capture); Conflicts over PAPs as replacements; Logic of freedom to compete	<u>Knowledge</u> In-house company knowledge dominates regulatory choices on product specification; Political usage of knowledge by e-NGOs to institutionally position themselves as experts and gain access to discussions over product specification

Table A1.2 Comparative sustainability interdependencies of Aquitaine trout

Fish farm/environment interactions

Social constructions of sustainability	Initial anthropocentric ecosystem approaches are increasingly challenged by quasi-ecocentric ecosystem approaches governing water quality		
Frontier politics of polity interdependencies	Territory French state-wide approach (to water policy) dominates; Conflicts between public actors' state-wide visions versus trout producers' local territorial visions of rivers	Regulation Sector-specific private self-regulation as counterpoint to dominant public rules	Knowledge Dominant regulative ecological science; Generation of self-monitoring data as potential leverage for new representations of impacts; Conflicts over interactions between farmed and wild trout

Access to fish farms

Social constructions of sustainability	Tensions between anthropocentric ecosystem approaches versus quasi-ecocentric ecosystem approaches. These tensions are opened up to controversy by producers and closed down to controversy by public officials in the Ministry of Ecology		
Frontier politics of polity interdependencies	Territory French-wide approach to water governance dominates in line with renewed French identity; Political work by producers in alliance with other actors to politicise differences over how different actors have connected visions of territorial futures with representations of the environment	Regulation Absence of strong public signal to grow the industry politicised through public–private mobilisation calling for new regulatory responses within ecosystem services' valuation; Tensions over nature as providing cultural services at the expense of economic activities	Knowledge Dominant regulative science; Technical knowledge gaps politicised but no destabilisation of hierarchies of knowledge

Sustainability as a food governance problem

Social constructions of sustainability	Institutionalisation of ecosystem approaches problematising ecology–society interdependencies and increasingly linked to sustainability of 'the homeland' have been complemented by social constructions of inter-relations between socio-ecosystems in non-contiguous places within a logic of equality		
Frontier politics of polity interdependencies	Territory No challenges to dominant EU political authority; Private rules define social meanings of Aquitaine's territories' sustainable trout as food (PGI Basque); Producers invest in work on local markets and local food supply chains; Territory as a value to compete for higher prices and as a weapon to protect home markets	Regulation Co-regulation governing sustainability marketing criteria on local environmental impacts; Reliance on global standards in self-regulation setting sustainability criteria for fish feeds; Conflicts over PAPs as replacements; Logic of equality	Knowledge Scientific expertise and in-house company knowledge support choices on product specification; Inter-professional body mobilises its acquired knowledge as expertise to legitimise its authority to set rules on product specification

Table A1.3 Comparative sustainability interdependencies of Greek seabass and seabream

Fish farm/environment interactions			
Social constructions of sustainability	Dominant de-politicisations of ecology–economy interdependencies are frequently attacked but not fully replaced by alternative constructions of sustainability, Seeds of alternative ecosystem approaches increasingly brought within the institution via implementation of EU rules		
Frontier politics of polity interdependencies	Territory Greek state-wide approach dominates and no political work to re-assign authority to another administrative scale despite criticisms	Regulation Public rules dominate and non-compliance has taken precedence over private self-regulation; New public-private consultation to renew and legitimise public rules	Knowledge Different knowledge types produced (scientific, producers, state) but conflicts over credibility of these different knowledges; Conflicts over environmental impacts

Access to fish farm sites			
Social constructions of sustainability	From de-politicised ecology–economy interdependencies to politicised anthropocentric ecosystem approaches Tensions over which economic activities should benefit nature and how to govern interactions		
Frontier politics of polity interdependencies	Territory Territory as a critical resource in conflicts between actors over how visions of coastal futures and representations of the environment have been aligned	Regulation Initial attacks on the sustainability institution through protests have been depoliticised through alteration of the frontiers between state and civil society; New form of participatory democracy institutionalised	Knowledge Scientification of controversies

Sustainability as a food governance problem			
Social constructions of sustainability	Initial depoliticisation of local ecology–economy interdependencies supplemented by depoliticisation of wider global inter-ecosystem effects		
Frontier politics of polity interdependencies	Territory Politicisation of contents of EU trade policy but not seeking to re-assign authority; Territory not explicitly marketed in selling of product; Equivalences sought between wild and farmed fish	Regulation Public rules and global standards sufficient for sustainability marketing criteria; Conflicts over PAPs as replacements	Knowledge Scientific expertise and in-house company knowledge to support marketing strategies

Summary tables around industry problems

Table A1.4 Around industry problems: farm/environment interactions

	ITD Terr + Nature/ Society	ITD Reg + Nature/ Society	ITD Know + Nature/ Society
Scottish Salmon			
Social constructions of sustainability	From de-politicised ecology–economy interdependencies to anthropocentric ecosystem approaches problematising dynamic ecology–society interdependencies		
Frontier politics of ITDs Territory; Regulation; Knowledge	Separation of Scottish approach and Scottish societal ownership; New identities and jurisdiction; Multi-positioning strategies on UK, EU scales	From private guidance on EIAs to public control; New public sectoral policy complemented by new private regulation	Claims of knowledge gaps to de-stabilise initial institutionalisation; Scientification of controversies in re-institutionalisation; Conflicts over interactions between farmed and wild salmon
Aquitaine trout			
Social constructions of sustainability	Initial anthropocentric ecosystem approaches are increasingly challenged by quasi-ecocentric ecosystem approaches governing water quality		
Frontier politics of ITDs Territory; Regulation; Knowledge	French state-wide approach (to water policy) dominates; Conflicts between public actors' state-wide visions versus trout producers' local territorial visions of rivers	Sector-specific private self-regulation as counterpoint to dominant public rules	Dominant regulative ecological science; Generation of self-monitoring data as potential leverage for new representations of impacts; Conflicts over interactions between farmed and wild trout
Greek seabass and seabream			
Social constructions of sustainability	Dominant de-politicisations of ecology–economy interdependencies are frequently attacked but not fully replaced by alternative constructions of sustainability. Seeds of alternative ecosystem approaches increasingly brought within the institution via implementation of EU rules		
Frontier politics of ITDs Territory; Regulation; Knowledge	Greek state-wide approach dominates and no political work to re-assign authority to another administrative scale despite criticisms	Public rules dominate and non-compliance has taken precedence over private self-regulation; New public-private consultation to renew and legitimise public rules	Different knowledge types produced (scientific, producers, state) but conflicts over credibility of these different knowledges; Conflicts over environmental impacts

Table A1.5 Around industry problems: access to farm sites

	ITD Terr + Nature/ Society	ITD Reg + Nature/ Society	ITD Know + Nature/ Society
Scottish Salmon			
Social constructions of sustainability	De-politicised ecology–economy interdependencies are replaced with anthropocentric ecosystem approaches. However, tensions emerge over hierarchy of ecosystem services being institutionalised		
Frontier politics of ITDs Territory; Regulation; Knowledge	Separation of Scottish approach; Political work by producers and SG to reduce conflicts over how different actors have connected visions of territorial futures with representations of the environment	From private planning to public control; Tensions between nature as providing multiple benefits to society versus nature as providing benefits to private economic actors	Scientification of controversies in re-institutionalisation
Aquitaine trout			
Social constructions of sustainability	Tensions between anthropocentric ecosystem approaches versus quasi-ecocentric ecosystem approaches. These tensions are opened up to controversy by producers and closed down to controversy by public officials in the Ministry of Ecology		
Frontier politics of ITDs Territory; Regulation; Knowledge	French-wide approach to water governance dominates in line with renewed French identity; Political work by producers in alliance with other actors to politicise differences over how different actors have connected visions of territorial futures with representations of the environment	Absence of strong public signal to grow the industry politicised through public-private mobilisation calling for new regulatory responses within ecosystem services' valuation; Tensions over nature as providing cultural services at the expense of economic activities	Dominant regulative science; Technical knowledge gaps politicised but no destabilisation of hierarchies of knowledge
Greek seabass and seabream			
Social constructions of sustainability	From de-politicised ecology–economy interdependencies to politicised anthropocentric ecosystem approaches. Tensions over which economic activities should benefit nature and how to govern interactions		
Frontier politics of ITDs Territory; Regulation; Knowledge	Territory as a critical resource in conflicts between actors over how visions of coastal futures and representations of the environment have been aligned	Initial attacks on the sustainability institution through protests have been depoliticised through alteration of the frontiers between state and civil society; New form of participatory democracy institutionalised	Scientification of controversies

Table A1.6 Around industry problems: farmed fish as food product

	ITD Terr + Nature/ Society	ITD Reg + Nature/ Society	ITD Know + Nature/ Society
Scottish Salmon			
Social constructions of sustainability	Attacks on sustainability as de-politicised local ecology–economy interdependencies lead to a re-institutionalisation implementing ecosystem approaches which recognise inter-relations between socio-ecosystems in non-contiguous places within a logic of competition		
Frontier politics of ITDs Territory; Regulation; Knowledge	No separation of Scottish public approach; EU public rules on food safety dominate; Private rules define social meanings of 'Scottish' sustainable salmon as food (PGI); Private actors engage in multi-positioning strategies; Territory as a value to compete for higher prices	Co-regulation governing sustainability marketing criteria on local environmental impacts; Self-regulation of sustainability criteria for fish feeds (policy capture); Conflicts over PAPs as replacements; Logic of freedom to compete	In-house company knowledge dominates regulatory choices on product specification; Political usage of knowledge by e-NGOs to institutionally position themselves as experts and gain access to discussions over product specification
Aquitaine trout			
Social constructions of sustainability	Institutionalisation of ecosystem approaches problematising ecology–society interdependencies and increasingly linked to sustainability of 'the homeland' have been complemented by social constructions of inter-relations between socio-ecosystems in non-contiguous places within a logic of equality		
Frontier politics of ITDs Territory; Regulation; Knowledge	No challenges to dominant EU political authority; Private rules define social meanings of Aquitaine's territories' sustainable trout as food (PGI Basque); Producers invest in work on local markets and local food supply chains; Territory as a value to compete for higher prices and as a weapon to protect home markets	Co-regulation governing sustainability marketing criteria on local environmental impacts; Reliance on global standards in self-regulation setting sustainability criteria for fish feeds; Conflicts over PAPs as replacements; Logic of equality	Scientific expertise and in-house company knowledge support choices on product specification; Inter-professional body mobilises its acquired knowledge as expertise to legitimise its authority to set rules on product specification
Greek seabass and seabream			
Social constructions of sustainability	Initial depoliticisation of local ecology–economy interdependencies supplemented by depoliticisation of wider global inter-ecosystem effects		
Frontier politics of ITDs Territory; Regulation; Knowledge	Politicisation of contents of EU trade policy but not seeking to re-assign authority; Territory not explicitly marketed in selling of product; Equivalences sought between wild and farmed fish	Public rules and global standards sufficient for sustainability marketing criteria; Conflicts over PAPs as replacements	Knowledge to support marketing policies

Index

Page numbers in **bold** denote tables.

adaptation 72, 74, 148, 171, 221
adaptive management 43, **49**
alternatives 8, 32, 47, 77–78, 84n14, 124, 156, 171; alternative interpretative frame 47, 112; alternative knowledge 72, 124, 182; alternative local economies 138, 143, 156, **211**; alternative policy directions 11, 81; alternative protein 4, 169; alternative society 47; alternative sustainability governing 3, 123, 127, 144, 197; alternative technologies 63
antagonistic politics 19, 209, 220

benefits of nature to society: and cultural benefits 150, 160; conflicts of interacting ecosystem services 153, 154, 221; diverse coastal communities 140, 141, 214; and equality 16, 138, 146, 151–154, 160; and spatial planning 154, 216
biophysical processes 33, 96, 170; and governing 66, 170; and innovation 99; and interactions with production processes 96, 101, 112, 124, 207; and landscape aesthetics 142, 149, 155; and 'naturalness' 14, 168, 172, 192, 194

cleavages 7, 45, 48
coastal communities 135, 138, 143–144, 153, 155; diverse coastal economies 214
compensation **82**, 128, 167, 200
consensus 7–8, 33, 78, 125, 209; relative consensus 3, 22n6
conservation 37, 38, 42–45, **49**; fish farming as conservation and restoration 4, 22n7, 98
contingency 11, 32, 67, 77, 81

convergence 220
co-regulation **82**, 183, 191

de-centralisation 9, 65–66, 215–216
dematerialised economies 47
democratisation of fish farming: of Aquitaine trout 215; of Greek seabass and seabream 217; of Scottish salmon 213
democratisation of knowledge 2, 9, 10, 60, 64, 75
demographic change 5
Drivers, Pressures, Impact, State, Response (DPSIR) policy tool 44, **49**

ecological continuity 98, 113, 125–126, 151, 216
ecological democracy 39, **49**
economic activity: and environmental impact assessments 218; interacting economic activities 159; and naturalisation 112; and society 76
economic instruments **49**
efficiency: as market theory 34–35; as political theory 4, 36–37, 48, **49**, **82**; production 47, 51n6, 99; as value 10, 48, **49**, 62
environmental degradation 2, 48–50, 63, 168; and degrowth 46–47; and ecosystem degradation 44; and human injustices 37; and technological fixing 34–36
environmental impact assessments 15, 35, 38–40, **49**, 50, 218; and Aquitaine trout 108, 110, 113–114; comparative usages 126–127; and Greek seabass and

seabream 116–118, 120, 122–123; and Scottish salmon 99, 104–105
equality: and competitiveness in Aquitaine trout 190, **191**, 200, 214; of nature's benefit to society 16, 138, 146, 151–154, 160; and Scottish salmon 214; and territory 5; as value 185

fish diseases: governing of in Aquitaine trout 111, 113, 125; governing of in Greek seabass and seabream 120; governing of in Scottish salmon 104, 105, 213; mitigation 139; spread of 5, 15, 97; treatment 87
fish health 5, 180, 202n24; and Aquitaine trout 111, 113, 215; comparative policy 126–127; EU directive **15**; and Greek seabass and seabream 117, 118; and Scottish salmon 102, 105, 107, 212
freshwater farming 1, 12, 15, 96, 98, 178
frontier politics 10–11, 76, 79–81, 208, 220; and knowledge 74–75; and regulation 69–71; and territory 66–67

global change 9, 33, 44, 66–67, 79, 221–222
governable, rendering sustainability 6, 11, 20–21, 139, 221
governor(s) 79, 77; credible governor 121; distant governors 125; dominant governor 119; economic governors 9; public actor governors 9
green constitutional rights 40, **49**, 70; constitutional commitment 38

institution, definition 7–8, 78, 207; *see also* institutional materiality
institutional and political work: definition 8, 76–80, 208; instrumentation 77; legitimation 77; problematisation 77
industry, definition 76–77
industrial production 7, 50n2
integrated management 37, **49**, 215; coastal zone management **15**, 137; ecosystem policy tools 43, 44, 216, 217; marine spatial planning 154–155, 157–159, 161

judicial activism 161, 219
justice, social 34, 36, 37, 39, 40, **49**

landscape 1, 14, **82**, 145, 146; aesthetic of landscape 52n17, 161; landscape

aesthetics 135, 141–144, 149, 155, 160; *see also* biophysical processes
legitimacy: and knowledge 74–75, **82**; procedural 39; and regulation 69–70, **82**; societal 65, 148, 153; and territory 65–66, **82**, 101; as value 10, 62, 66
liberal capitalism 34; liberal capitalist state 9, 39; *see also* neoliberalisation
linear models: of change 220; of expertise 71, 84n7; of governance 8, 60; of nature 43
lobbying 114, **211**, 217, 219
local economy 156, 186
local employment 5, 162n15, 184

management tools 35, 43, 136–138
material: empirical 12, 17, 20, 32, 79, 93; institutional materiality 10, 101; raw materials for fish feeds 177, 181, 182, 189, 198; *see also* dematerialised economies
mitigate 6, 218; mitigation 82, 124, 126

neoliberalisation 9, 68; neoliberal governance 214; neoliberal growth 47; neoliberal industrial capitalist system 46; *see also* liberal capitalism
nutrition: fish 1, 171, 172; human 6; political use of nutritional knowledge 182, 190; processed animal proteins and fish nutrition 176, 181, 189, 191

ocean health 1, 16, 31
opponents 5, 7

polemical debates 21; polemical discussions 3
policy–polity tensions 2, 20, 59, 61–62, 81, 208
political economy, tangled politics 219; *see also* tangled politics
political modernisation 62, 64, 71
political sociology: and institutions 17, 60–61, 76, 81; and politics 10
politicisation, definition 77
pollution: defined as a public problem 33, 111; and regulation 35, 83n4, 102, 105; and territorial interdependence 66
precautionary principle 39, **49**
Processed Animal Proteins (PAPs) 171–172; and Aquitaine trout 189; and Greek seabass and seabream 196; and Scottish salmon 176, 180–181

public action principle, sustainability as institution 7–8, 76, 78, 207
public policy analysis 9, 61

raw materials for fish feeds 177, 181, 182, 189, 198
redistribution of power 9, 48, 68, 79, 212
redistributive policies 64–65
regionalisation 2, 65; and Aquitaine 110; and Greece 117; and Scotland 101
resources: EU policy 119, 149, 155; fish 3, 4; institutional 63, 65, 80, **82**, 101, 209; marine 4, 171; natural 16, 33, 42, **82**, 122; socio-economic 67, 110; territorial 9, 141, 175, 220; water 13, 109
restoration 22n7, 42, 98, 200n2, 216
re-wilding 42, **49**
rural economies 1, 140

seawater farming 12, 13, 15, 96
sectoral policy strategy 100
smart regulation 21n2, 35, 49
social welfare 36, 40
sovereignty 9, 64, 214
sustainability social contract 39

tangled politics 82, 125, 208–210, 219, 222; tangled processes of institutionalisation 93; *see also* political economy
temporality: temporal accounts 18; temporal processes 8; temporal scale 44, 61, 124

tourism: and growth 135, 154; integrated with fish farming 141, 146, 154, 157, 158, 187; and spatial planning 153–156, 156, 159, 216; and sustainability 222n1; versus fish farming 5, 16, 153–156, 221

utilitarian politics 48, **49**, **82**, 83, **211**, 212; and comparative aquaculture policies 220–221; and ecosystem politics 45, 221

values: consumer 171; economic 45; governing 62; political work 77; scientific versus societal 71; sustainability 8, 32, 39, 50; *see also* efficiency; equality
visual impact assessment 145–146, 162n11, 200

welfare: animal 1, 103, 119, 128n18, 195–196; social 36, 40
Westphalian state 9; transformation 64
wild fish: compared with farmed fish 192, 194, 198; fish farming creating pressures on wild fish stocks 5, 167–168; fish farming reducing pressures on wild fish stocks 16, 45, 136, 176; interactions with farmed fish 5, 97–98, 124; interactions with salmon 105–106, 113, 124, 162n16; interactions with trout 125
working groups 100–102, 213; and containment 105–106; and feeds 175, 176, 201n13; and fish health 105–106; and planning 145, 146, 161

Printed in the United States
by Baker & Taylor Publisher Services

Printed in the United States
by Baker & Taylor Publisher Services